Science and Fiction

Science and Fiction – A Springer Series

This collection of entertaining and thought-provoking books will appeal equally to science buffs, scientists and science-fiction fans. It was born out of the recognition that scientific discovery and the creation of plausible fictional scenarios are often two sides of the same coin. Each relies on an understanding of the way the world works, coupled with the imaginative ability to invent new or alternative explanations—and even other worlds. Authored by practicing scientists as well as writers of hard science fiction, these books explore and exploit the borderlands between accepted science and its fictional counterpart. Uncovering mutual influences, promoting fruitful interaction, narrating and analyzing fictional scenarios, together they serve as a reaction vessel for inspired new ideas in science, technology, and beyond.

Whether fiction, fact, or forever undecidable: the Springer Series "Science and Fiction" intends to go where no one has gone before!

Its largely non-technical books take several different approaches. Journey with their authors as they

- Indulge in science speculation – describing intriguing, plausible yet unproven ideas;
- Exploit science fiction for educational purposes and as a means of promoting critical thinking;
- Explore the interplay of science and science fiction – throughout the history of the genre and looking ahead;
- Delve into related topics including, but not limited to: science as a creative process, the limits of science, interplay of literature and knowledge;
- Tell fictional short stories built around well-defined scientific ideas, with a supplement summarizing the science underlying the plot.

Readers can look forward to a broad range of topics, as intriguing as they are important. Here just a few by way of illustration:

- Time travel, superluminal travel, wormholes, teleportation
- Extraterrestrial intelligence and alien civilizations
- Artificial intelligence, planetary brains, the universe as a computer, simulated worlds
- Non-anthropocentric viewpoints
- Synthetic biology, genetic engineering, developing nanotechnologies
- Eco/infrastructure/meteorite-impact disaster scenarios
- Future scenarios, transhumanism, posthumanism, intelligence explosion
- Virtual worlds, cyberspace dramas
- Consciousness and mind manipulation

More information about this series at http://www.springer.com/series/11657

Damien Broderick

The Time Machine Hypothesis

Extreme Science Meets Science Fiction

Springer

Damien Broderick
San Antonio, TX, USA

ISSN 2197-1188   ISSN 2197-1196   (electronic)
Science and Fiction
ISBN 978-3-030-16177-4   ISBN 978-3-030-16178-1   (eBook)
https://doi.org/10.1007/978-3-030-16178-1

This Springer imprint is published by the registered company Springer Nature Switzerland AG
The registered company address is: Gewerbestrasse 11, 6330 Cham, Switzerland

*For Kat and Ryan
and Aurelia and Charlotte,
my step-persons
And as always for dear Barbara,
my beloved person-person*

Acknowledgments

I am delighted to thank my dear wife, Barbara Owsley Lamar, for her love and support as I researched and wrote this exploration of time machine science and fiction. Also my time traveling California pal Gary Livick, whom I have never met although I have seen pictures of his two dogs, for emailed advice and very silly jokes.

Thanks as well to Angela Lahee, my commissioning editor at Springer, and her associate Rebecca Sauter, who have kept the wheels turning on this and my earlier volume (*Consciousness and Science Fiction*) in the Springer *Science and Fiction* series. I am grateful to Professor Gregory Benford, scientist, novelist, and editorial board member of that series, and to Dave Truesdale for useful referee suggestions.

In general, I applaud the intelligence and bravery of the many daring scientists whose books and papers have guided my quest for a plausible time machine hypothesis and the novelists and short story writers whose science fiction led me along so many intriguing pathways in the corridors of time.

Today the subject of time travel has jumped from the pages of science fiction to the pages of physics journals as physicists explore whether it might be allowed by physical laws and even if it holds the key to how the universe began. In Isaac Newton's universe time travel was inconceivable. But in Einstein's universe it has become a real possibility.... To appreciate what scientists are studying now, an excellent first step is to explore major time-travel themes in science fiction, where many ideas in this arena were first advanced.

J. Richard Gott, III, Princeton University emeritus professor of astrophysical sciences (*Time Travel in Einstein's Universe,* 2002, 5)

Contents

Part I

Spacetime Time

1

Time Travel Unraveled

…to speculate openly about time travel is tricky. If the press picked up that the government was funding research into time travel, there would either be an outcry at the waste of public money, or a demand that the research be classified for military purposes. After all, how could we protect ourselves if the Russians or Chinese had time travel and we didn't? They could bring back Comrades Stalin and Mao! So there are only a few of us who are foolhardy enough to work on the subject that is so politically incorrect, even in physics circles. We disguise what we are doing by using technical terms like "closed timelike curves," which is just code for time travel.

Professor Stephen W. Hawking, "Chronology Protection" (2002, 87)

It may interest you to know that there are real physicists out there who are busily publishing papers on whether and how a practical time machine can be built.*

*Rest assured, they already have tenure. (*A User's Guide to the Universe*, 2010, 154–55)

Shortly before the turn of the twentieth century, in 1895, the first time machine story was published by Herbert George Wells, titled (I'm sure you've guessed) *The Time Machine*. Science fiction, as a distinct commercial and literary means of speculative story-telling, has a long pre-history, at least from Mary Shelley's *Frankenstein*, 1818, to Mark Twain's *A Connecticut Yankee in King Arthur's Court*, 1889). Even so, Wells can be seen fairly as the initiator of what we now regard as a fresh genre, or—better still—fresh narrative mode.

It is perhaps significant that his great leap into true novelty was an exploration of deep time, immense futurity, Darwinian evolution via environmental selection. Later he pioneered rather less extreme topics such as invisibility and

© Springer Nature Switzerland AG 2019
D. Broderick, *The Time Machine Hypothesis*, Science and Fiction,
https://doi.org/10.1007/978-3-030-16178-1_1

even the horrifically plausible: global war (against Martians and their technology, a metaphor for German might), military tanks, atomic weapons, powered bombers, a "General Intelligence Machine" or AI, landing on the Moon (although via anti-gravity, which might yet prove valid in this era of Dark Energy), enhanced biology uplifting animals into conscious humanoid slaves. But his deepest resonance proved to be the vehicle carrying its unnamed narrator into the future, ultimately all the way to Earth's expiration date under a dying Sun.

Why time travel? Perhaps as an imaginative response to the vast gulfs of time revealed by nineteenth century geology and biology, the profound shocks of heredity shaped by natural selection rather than divine forethought, the vast age of Earth and cosmos alike. When steam trains came into being because of convergent technologies and aspirations, it was "Steam Engine Time." For the brilliant Wells, the dazzling light ahead was, in 1895, a kind of metaphysical and ultimately engineering prospect of machine command over time. By the end of his journey the light had dimmed with the price paid, the ultimate in future shock. Even so, Wells was drawn, like many of his admirers in the following century, into the echoing, dangerous vaults and distances of Time Machine Time.

But wait! Despite Wells's fiction, and what Richard Gott and Stephen Hawking claimed, isn't time in actuality a one-way street, straight from the unchangeable past to the still-unborn future? So how could a time machine possibly *work*? What would a real-world time machine *be*?

Here is a disheartening example of what it would *not* be. In an overwrought and unjustifiably confident book, *Breaking the Time Barrier* (2005), the British journalist Jenny Randles reported a supposed invisible "unseen fifth dimension" explored by the USAF using a "noctovision camera" (72)—that is, a night vision camera designed to be especially sensitive to infrared frequencies.

A miracle happened!

> Mounted aboard a high-flying aircraft, the camera recorded infrared energy from a parking lot thousands of feet below. The parking lot was completely empty… but *inexplicably* [my emphasis, after shaking my head and rubbing my eyes] the photographs showed the ghostly images of whole lines of parked cars… Were they in another dimension? (73).

Well, no, of course they weren't. At the end of work hours, the vehicles had been driven away after a day's hot Florida sunlight baked the tarmac, except where each car had been parked. Hot tarmac unprotected by the shade under parked cars would register strongly on the film for some hours, leaving cooler "footprints" in the still-radiating surface. Naturally, a camera designed to detect infrared (that is, heat) images showed the vehicles' lingering heat shadows.

Randles makes this sound transcendental, like the faithful finding the face of Jesus burned into a piece of toast. "The energy emissions" (that is, the heat) "detectable from the cars before they departed could be used to recreate the residual outlines of their body shapes" (73). In reality, it would be the surrounding concrete or asphalt surface that would paint its impression inside the camera.

The same goofy story had been published 39 years earlier in *Beyond the Time Barrier*, by Russian-Australian Andrew Tomas (1906–2001). In turn, this version of what Tomas called "earth-shaking experiments" was sourced in an Associated Press report from August 17, 1958, found in the *Miami Herald*. "The officials did not release their 'time pictures' to the public for fear that trained scientists in other countries might have 'read too much from them.'"

It ended with a flourish rivaling Ms. Randles': "To take a snapshot of empty space and see one's car on the photograph at a spot where it was two hours ago, is truly breaking the barrier of time" (51–2).

This is flagrant misrepresentation, the kind of silliness that gives any possibility of real time machines an even worse reputation.

We need to start with a simple, non-bogus definition. (By and by, there will be more complicated varieties.) For now, let's just propose that a time machine is a hypothetical device or condition of the cosmos able to propel a live passenger, or an instrument package, rapidly into the past or future. Preferably, it should be capable of returning that passenger or package to its time and place of departure.

Hence, we are faced immediately with a startling question. If time travel is feasible, not a breach of the established laws of physics, is it possible that we are already being observed by visitors from the future or even the past?

—Visitors who have travelled back or forward to our era in time machines yet to be invented or even from forgotten civilizations (whether human or alien) lost in the mists of prehistory?

—Maybe even from *alternative* histories, where things have turned out differently?

Answer One: Could be, according to some scientists of the caliber of Stephen Hawking, but not likely.

Answer Two: No! Time travel, like perpetual motion and the Flat Earth delusion, is out of the question. Absolute nonsense. Get help.

Answer Three: Why, certainly it's possible. Just the other day, I was chatting to a very nice person from the thirty-first century.

Most intelligent people, I guess, would have no hesitation in hitting the buzzer for Answer Two. Wells's inventive conceit, and *Back to the Future, The Terminator, Groundhog Day*, are acceptable as fun fantasy. Yet proposing that time travel might someday be a reality seems ridiculous. Taking it seriously, except as an exercise in playful imagination or science fiction speculation, would be literally childish. Maybe even certifiably crazy.

Until some decades ago, hardly a scientist in the world would have given any reply but Answer Two: "Absolute nonsense. Get help."

Then science hit a hump in the road.

Weird and improbable things came pouring out of seemingly innocent (if often maddeningly complicated) equations. To start with, as far back as the early days of Albert Einstein's General Relativity, there was the possibility of black holes. These collapsed stars would be more massive than the Sun, even immensely more massive, their gravity so concentrated into a small compressed space that not even light could escape its grip. Here physics reached its limits, in the absence of a solid theory of quantum gravity.

For a while, physicists (including Einstein) tightened their seat belts and waited for black holes to go away, hoping they were just figments of the imagination that would vanish when calculations were done more carefully and in greater detail. Instead, to the dismay of conservative experts, the physical evidence for black holes scattered throughout the universe firmed with every year of additional research. Now we are almost certain there's a gigantic one right in the dusty center of our Milky Way galaxy, and hundreds of billions more throughout the universe.[1] They might be more numerous than the visible stars. We shall look in more details at this and related topics shortly, since they form a convenient entry point to current investigations into time travel.

In 1965, Professor Richard Phillips Feynman shared the Nobel Prize for physics. At the center of the profound mathematical studies for which he was honored stood an extraordinary conjecture on time reversal. It had been published 16 years earlier in *Physical Review*, as "The Theory of Positrons." The genesis of that conjecture was detailed drolly in his Nobel acceptance lecture, "The Development of the Space-Time View of Quantum Electrodynamics."

Working towards a Ph.D. at Princeton, while in his early twenties, Feynman got a phone call from his colleague J. A. Wheeler, even then a notable theoretician.

"Feynman," Wheeler told him, "I know why all electrons have the same charge and the same mass."

---

[1] https://www.nytimes.com/2018/10/30/science/black-hole-milky-way.html

This was (and still is) a considerable mystery to physicists. Throughout the known universe, as we learn from the sophisticated instruments of astronomers, elementary particles such as the negatively charged electron appear to be basically identical. Reasonable enough, at first sight.

But then no other phenomenon on a larger scale is so homogeneous. Trees of a given kind are similar, but hardly identical in root and branch. There are many stars akin to the Sun, but no two exactly alike. Even crystals differ. Yet electrons, aside from variations in their energies, are indeed strictly indistinguishable one from the next.

Feynman was intrigued. "Why?" he asked.

"Because," Wheeler told him triumphantly, "they are all the same electron!"

It was an audacious jest. What Wheeler was suggesting was simple—and mind-wrenching.

Suppose an electron, moving forward in time along with the rest of us, got kicked into the past by a powerful electromagnetic field? Then it would appear to us, earlier, as a positron, a positively-charged antiparticle identical to an electron but opposite in all detectable electromagnetic effects. Still earlier, in our terms, plunging into the past, the positron might be hurled forward again, by an energetic exchange, along the "normal" direction of time. Now the signs would be reversed once more, and the particle would manifest itself to us as a second electron elsewhere in space.

Repeat this process an indefinite number of times, and you can account for every electron in the universe. They are all the same electron, weaving and shuttling across the entire history of the universe from the Big Bang at its origin to its unknown, distant death.

The concept is breathtaking.

Wheeler was joking, of course, but the notion exhilarated Feynman. In his 1965 Nobel lecture, he said: "I did not take the idea that all electrons were the same one from him as seriously as the observation that positrons could simply be represented as electrons going from the future to the past in a back section of their world lines. That, I stole."

Wheeler's jest proved immensely fruitful. By 1949, Feynman was able to show that the explosive meeting of a positron and an electron (in which they annihilate each another in a burst of energy) is exactly the same as a catastrophic time reversal on the part of the electron alone.

Similarly, a photon (or quantum of pure energy) can decay into an electron-positron pair, and this event is indistinguishable from the picture of a backward-moving electron (seen by us as a positron) being kicked forward again into "normal" duration by its impact with the photon.

Many physicists today treat Feynman's discovery as nothing more than a highly useful book-keeping device, a graphic way to keep track of the interchanges of energy between particles. But it remains true that within certain boundary conditions we are justified in interpreting anti-particles such as the positron as genuine time travelers.

Certainly they're not the kind celebrated in science fiction stories, but the fact remains that for more than 70 years orthodox physics has had to deal with the possibility of time reversal.

Feynman's demonstration was far from easy to swallow. In 1956, in *The Direction of Time*, Hans Reichenbach acknowledged its significance by dubbing it "the most serious blow the concept of time has ever received in physics." Since then, even more astounding concepts have emerged in the symbols of mathematicians. But it is arguable Feynman's breakthrough was the moment when the Time Machine Hypothesis first became intellectually conceivable.

Worse than black holes was the theoretical possibility of traversable wormholes (hyperspace shortcuts between two points very distant in spacetime). This possibility revealed by the renovated physics was dubbed the Einstein-Rosen bridge, or wormhole, an idea linked with Professor Wheeler, Feynman's inventive friend. Wormholes are conceived as joining points in space and time through a higher dimensional realm, dubbed Superspace in some interpretations and Bulk in more recent models. Can we grasp what this foamy, wormhole-filled realm that pervades spacetime is like? Not readily. Reaching for a way to convey this marvel, Wheel expressed himself in a kind of poetry:

> It is like chasing after Merlin. One moment it is a rabbit, and the next a gazelle. And just as you reach out to touch it, it turns into a fox, or a brightly-colored bird fluttering on your shoulder. It is the place where smoke comes out of the computer because all the classical laws of spacetime break down… In Superspace the question "what happens next?" is devoid of content. The very words "before," "after" and "next" have lost all meaning. Use of the word "time" in any normal sense is completely out of the question.

So mass might disappear into a wormhole and emerge far away. "If the mass passes through the wormhole and suddenly appears a billion light-years away in ordinary space once more," Dr. Isaac Asimov commented soon after they were postulated, "something must balance that great transfer in distance.

Apparently this impossibly rapid passage through space is balanced by a compensating passage through time, so that it appears one billion years ago."[2]

No nearly instantaneous wormhole warp-gates have yet been found in space nor created in the lab. But it is clear that these issues remain open and vital. For if wormholes truly exist, they spell greater disruption than black holes do for the classical picture of space and time. For example, if a wormhole is built or discovered in plain sight, its ends (braced open by "exotic matter"—negative energy, negative mass) could form a bridge from one time to another, as described in considerable detail in scientific papers[3] and careful popular treatments of the notion by both Kip Thorne (*Black Holes and Time Warps*) and Richard Gott (*Time Travel in Einstein's Universe*).

The simplest way to do this (where the word "simplest" is really a grotesque irony) would be to hold one end steady while carrying the other end into space at close to the speed of light, or spinning around a black hole, to retard its rate of change. Then carry the wormhole mouth back to your starting point. Throughout this expensive and improbably difficult maneuver, both ends of the wormhole will remain in contact and share the same metric frame, which means that you could step through one portal and emerge from the other into the past or future. The major physics problem here, as Stephen Hawking claimed, is that because the wormhole bridge is an open channel, stray particles and even vacuum energy "virtual particles" would wander through and back, gaining energy with each circuit. Very swiftly, this gale of self-amplified raw energy (like the deafening shriek of positive feedback from two adjacent microphones) would rip the spacetime of the wormhole asunder, along with anyone trying to pass through it.

Is there any way to evade this horror? It might be feasible, in principle at least, to brace each end of the wormhole with "exotic matter." If it exists, can such stuff be harvested and worked into a suitable framework to save wormhole passengers from instant obliteration? This remains a question for theorists and future experimenters. But it must be admitted, the prospects of a traversable, stable wormhole do seem unlikely at the moment.

So does the reality of tachyons, a postulated class of sub-atomic particles favored by science fiction since their existence was postulated. If they turn out to exist, tachyons must travel *always* faster than light, and hence backwards in time.

---

[2] Isaac Asimov, *The Collapsing Universe* (1977), 219.
[3] See the partial list of peer-reviewed papers by these scientists and their colleagues, at the end of this volume.

To date, the querulous may take comfort in the failure of experimentalists to catch any tachyons in their laboratories during the last half century or so, even though some Large Hadron Collider specialists at CERN thought for a giddy moment, in 2011, that neutrinos might be tachyons. Alas, it was just a faulty diode in an optic fiber link, producing an erroneous reading.

Still, in light of these early investigations the feasibility of manipulating time began to seem less outrageous,. The brilliant Austrian logician and mathematician Kurt Gödel, a friend of Einstein's at the Institute for Advanced Study in Princeton, found that the equations of General Relativity permitted time travel to the past in a universe that is rotating. The rebuttal came quickly that our universe is *not* rotating, which can be proved. But this does not invalidate the worth of Gödel's mathematical investigation, which provided a door for us to keep ajar. Gott notes:

> …the Gödel solution is very important, for it showed that time travel to the past is possible in principle, with Einstein's theory of gravity. If there is one solution that has this property, there can be others. (92)

Finally, and most controversially (you may close your eyes and hold your nose here if your stomach isn't up to it), the offensive claims of laboratory psychic research crept from the shadowy borderlands on the outskirts of physics and psychology, and stood on the doorstep of the Academy. Some prestigious scientific journals, reluctantly and with grave cautions, found it necessary to publish papers on retrocausal effects—where the future influences the past to some extent—and phenomena dubbed "anomalous cognition" (knowing the future in advance) and perhaps "anomalous perturbation" (intention operating directly on matter).

These phenomena presented faint but cumulatively impressive evidence that the human mind itself can breach the accepted limitations of entropic space and time. See, for example, the recent paper by Professor Etzel Cardeña (2018): "The experimental evidence for parapsychological phenomena: A review." This appeared in the quite straitlaced and important mainstream journal *American Psychologist*.[4]

Allegedly (indeed, demonstrably[5]), some people can give statistically impressive double-blind descriptions of remote locations, unknown to them by normal means, while being monitored in laboratory conditions. Even

---

[4] https://doi.org/10.1037/amp0000236

[5] See, for example, three recent academic collections: Broderick and Goertzel, eds., *Evidence for Psi* (McFarland 2015), May and Mahawa, eds., *Extrasensory Perception*, 2 volumes (Praeger 2015), and Cardeña, Palmer and Marcusson-Clavertz, eds., *Parapsychology: A Handbook for the 21st Century*

more drastic, such "remote viewing" worked just as well if the randomized, hidden target location was not chosen until *after* the viewing session was complete. That is, these gifted people possessed, to some more-than-chance degree at least, the ability to foresee the unpredictable future. If this claim continues to hold up, then these remote viewers are not *using* time machines. They *are* time machines, neurologically speaking. Spooky, I know.

Still, even if we restrict our attention for the moment to black holes and wormholes, it is abruptly clear that the only fair reply in the 2020s to the question of time travelers will be Answer One: *Maybe such devices are possible, even if not for the likes of us twenty-first century humans.*

Time machines, in short, are no longer only a favored fancy of science fiction authors like me. Whether or not they literally exist is still uncertain, but their status has shifted from that of playful and inventive fantasy to the respectable realm of genuine scientific speculation. Consider the following notable science specialists (and academically trained science popularizers), several already cited above, who in the last couple of decades have spent time investigating the possibilities and paradoxes of travel through time. It seems appropriate to track their entrance into this topical domain in order of publication.

Stephen Hawking was born in 1942 and died in 2018 after decades of increasing physical paralysis from motor neuron disease. For 30 years he was Lucasian Professor at Cambridge, holding the chair held by Isaac Newton 300 years earlier. His *A Brief History of Time* became a famously best-selling if often unread book in 1988. In its updated edition a decade later, he introduced an entire new chapter on the topic of time travel, based on formal papers published earlier that helped kick-start the academic interest in moving freely back and forth in time.

Paul J. Nahin, born in 1940, took a PhD from the University of California, Irvine, in 1972, and is currently an emeritus professor of electrical engineering at the University of New Hampshire. A quarter century ago, he published a hefty treatise on both the science and science fiction of time travel prospects. *Time Machines: Time Travel in Physics, Metaphysics, and Science Fiction* was published, gratifyingly, by the American Institute of Physics. It reappeared from Springer in a revised and updated 2017 version, *Time Machine Tales: The Science Fiction Adventures and Philosophical Puzzles of Time Travel.*

(McFarland 2015), plus the long-classified documentation from the US government psi program, May and Mahawa, eds., *The Star Gate Archives: Remote Viewing*, vols 1 and 2 of 4 (McFarland 2018).

One of the finest and most inventive studies in our topic is *Time Travel In Einstein's Universe: The Physical Possibilities of Travel through Time* (2001) by J. Richard Gott, born 1947, a professor of astrophysics at Princeton.

An equally clear, short examination of our topic is British polymath Paul Davies's *How To Build a Time Machine* (2001). In July 2018, BBC television ran a Horizon program on the topic borrowing (or plagiarizing) the same title. Davies, born 1946 in London, emigrated first to Australia and then to the US, is now a research professor at Arizona State University, pursuing topics in cosmology, quantum field theory, and most recently astrophysics. He has won a lucrative Templeton Prize, and the Royal Society's Faraday prize, and in 2007 was granted membership in the Order of Australia, a sort of knighthood in an egalitarian culture. He also has an asteroid named after him: 6870 Pauldavies.

Ronald Mallet (born 1945) became a full professor at the University of Connecticut in 1987, specializing in cosmology and astrophysics. In *Time Traveler: A Scientist's Personal Mission to Make Time Travel a Reality* (2006), he proposed methods to travel into the past by creating relativistic frame-dragging (in which accelerated rotation can cause spacetime to literally twist in the vicinity), and even closed timelike curves.

Sean Carroll, born 1966, is another cosmologist and quantum physicist, a research professor in the Dept. of Physics at the California Institute of Technology. His current interests are primarily in general relativity and the mysteries of dark energy. A rigorous but highly imaginative skeptic, he makes it clear in *From Eternity to Here: The Quest for the Ultimate Theory of Time* (2010) that time travel is not impossible, just very, very difficult, based on what we know today.

Perhaps the most remarkable and up-to-date treatment of wormholes, plus gravity used as a time portal driver, and the marvels of bulk and brane (we'll return to these rather difficult multi-dimensional regions) is by Nobel Laureate Kip Thorne, born 1940 and before his 2009 retirement Feynman Professor of Theoretical Physics at the California Institute of Technology. His earlier marvelous book was *Black Holes and Time Warps* (1994). Perhaps he is most famous these days as the theoretician of wormhole travel in the novel and subsequent movie *Contact* (1985, 1997), and recently the more rigorous film *Interstellar* which he explains as a cautiously scientific interpreter in *The Science of* Interstellar (2014).

This list omits other books by science writers not specifically involved in time travel research, such as Clifford A. Pickover's *Time: A Traveler's Guide*, released by Oxford University Press, in 1998. Pickover, somewhat like Isaac Asimov, is a polymathic popularizer of science. Born in 1957, with a 1982

PhD in molecular biophysics from Yale, he has some 500 patents—although so far none in time travel dynamics. Brian Clegg, born 1955, holds two Master's Degrees, and writes popular science. His 2011 book has the same title as Paul Davies'—*How To Build a Time Machine*, but the 2012 British edition carefully retitled it as *Build Your Own Time Machine*. Neither of these books actually contains a blueprint for achieving this self-assembly, but they do convey the general drift of the science of time travel.

Given this barrage of research and publications, it is perhaps timely that a biographer of Feynman (*Genius*) James Gleick (born 1954), released a sort of nostalgic study of these recurrent themes in *Time Travel: A History* (2016). Not himself a researcher, Gleick insightfully traces the paths taken by these earlier explorers and leaves us at the intersection where dreams and advanced physics have grown entangled:

> Every few years someone makes headlines by hailing the possibility of time travel through a wormhole—a traversable wormhole, or maybe even a "macroscopic ultrastatic spherically-symmetrical long-throated traversable wormhole." I believe that these physicists have been unwittingly conditioned by a century of science fiction. They've read the same stories, grown up in the same culture as the rest of us. Time travel is in their bones." (58)

To our considerable surprise, therefore, we can now see that time travel might be at least *possible* within the terms of advanced current scientific theory. Shortly, I shall detail some of the ways that might be.

Then we can explore the many paths along which this notion has wandered in science fiction.

Finally, we shall look at some of the (questionable) evidence suggesting time travel is not just *allowed* by physics, but indeed might be a reality that intersects our present-day world, offering threats and promises more extraordinary than the prospect of travel to the stars.

As we shall see, the Time Machine Hypothesis has no lack of theoretical support. It has yet to confront formidable objections, however, far graver than those which stood for so long in the way of plate tectonics, once dubbed "drifting continents." And the hard evidence in its favor is scanty. The notion is still so new, historically, so challenging, that no considerable search for evidence has been undertaken. Regard the comments on the possibility by Hawking, from the additional chapter on time travel and wormholes in the second edition of his landmark book:

We thus have experimental evidence both that space-time can be warped… and that it can be curved in the way necessary to allow time travel…. One might hope therefore that as we advance in science and technology, we would eventually managed to build a time machine. But if so, why hasn't anyone come back from the future and told us how to do it? There might be good reasons why it would be unwise to give us the secret of time travel at our present primitive state of development, but unless human nature changes radically, it is difficult to believe that some visitor from the future wouldn't spill the beans. Of course, some people would claim that sightings of UFOs are evidence that we are being visited either by aliens or by people form the future. (If the aliens were to get here in reasonable time, they would need faster-than-light travel, so the two possibilities may be equivalent.

Stephen Hawking, *A Brief History of Time,* 2nd edition, 1998, pp. 165–6.

As Hawking implied, the most dramatic and incontrovertible proof would be the sighting of a time machine by credible witnesses. At once we are brought into the irritating zone indicated by Answer Three ("Some of my best friends are from the far future!"). I've deliberately pitched Answer Three at a whimsical level. Our instant reaction to a claim that someone has witnessed a time machine, let alone traveled in one, would be to roll our eyes, if not jeer.

I present the Time Machine Hypothesis here as entertainment, as speculation, an exercise in following both current science and science fiction into the realm of the (barely) possible. If it is valid, though, it implies that future civilizations will one day master time travel. If so, it is not an absolute absurdity that even now they might be among us, their ancestors, observing our quaint and ferocious ways.

2

The Scientific Basis for Time Machines

Space and time are interchangeable (except for our inability, so far, to double back chronologically as we can spatially). So Einstein told us. An interval that one observer sees as a gap in space between two places can be seen just as validly as a gap in time between two moments. It is a truly shocking discovery.

By now it is well known that a space traveler flying to a distant star system at speeds closely approaching that of light will return home to find that his 10 year voyage has been experienced on Earth as a hundred or a thousand years.

He has experienced time dilation. His great speed has shifted him to a different frame of reference, causing time on his craft to run slow by comparison with clocks on Earth.

What's seldom grasped is that space, for the traveler, has actually been compressed or contracted. Say his 10 year round-trip was to a star known to be 50 light years distant. That's a total journey of 100 light years, authentically experienced by him as lasting just 10 years. So it seems he must have travelled ten times faster than light. Yet Einstein insists that ordinary matter cannot be accelerated faster than light.

The paradox vanishes when we grasp that for the voyager, due to space contraction, the distance between Earth and his destination has shrunk (literally) to less than five light years. At his enormous speed, the universe flattens in front of him. For the crew of that starship, all of reality squeezes together. The universe bunches up.

In fact, in the overall single reality of spacetime, the balance between time and space shifts according to which observer we question. To the earthbound unaccelerated witness, it appears that time on the spacecraft slowed down. To the space traveler, it appears that the distance covered shrank.

© Springer Nature Switzerland AG 2019
D. Broderick, *The Time Machine Hypothesis*, Science and Fiction,
https://doi.org/10.1007/978-3-030-16178-1_2

Both interpretations are equally valid, because, in a sense, space and time dimensions can be transformed into one another. It is the combination of both time and space into a single spacetime value that creates the *invariant* reality expressed from each observer's standpoint. This is a little like the way 2 × 9 = 18 yields the same combined result as 3 × 6, also equaling 18.

Time dilation under acceleration sounds impossible—even logically incoherent. But there is overwhelming experimental evidence that this understanding of reality is correct. No other view to date makes as much sense of the universe, from the most elementary subatomic particle to the most remote quasar, as does this fantastic perspective.

So much for common sense.

Such a hammer blow to the traditional wisdom of humankind is not easily absorbed, even after a century. The truth of Relativity has been accepted by scientists for more than four generations, but the deep implications of the news have not really seeped through to the non-scientific public. It's doubtful, for that matter, whether it has been grasped emotionally by most physicists. Yet the interchangeability of time and space is just the first crack in the foundations of common sense. The towers of our ancient certainties are eroding fast. For if space and time can be swapped, so that a year aboard a spacecraft travelling at 99.9996% of light speed becomes a day and a kilometer shrinks to two and three-quarter meters, might not time travel itself become feasible? Could our relentless motion forward in time be not just slowed but reversed, just as we can retrace our steps through space? That question, however, is naïve, and fails to understand that time *stops* at the speed of light in vacuum; it does not *bounce*, or begin running backward.

For all that, though, it turns out that there is nothing in the principles of General Relativity, Einstein's theory of gravity, to deny the possibility of retro-causal effects. And there is much evidence to support its reality.

Advanced science sometimes has the look of a banana republic. Revolutions sweep across the terrain with dizzying regularity, toppling old regimes and enthroning young upstarts who fall in their turn before the decade is out. Unlike political revolutions, though, the turmoil is fairly bloodless, and all the combatants benefit. In the 1970s and 1980s, the innovation that captured most public attention was the likelihood of black holes in space. It was swiftly realized out that black holes might be a gateway into time travel: that the black hole concept straddles two regimes of science.

"Bizarre though they may seem in everyday terms," wrote Dr. John Gribbin, 40 years ago, "an astronomer brought up in the old school can, perhaps with a little discomfort, stretch his mind to accommodate the concept… But, of course, the old astronomy is not the ultimate world picture… These objects may be the most peculiar feature of the old astronomy, but they are also among the simplest and most 'obvious' features of the new astronomy." Some doubts lingered for a time as to the physical reality of black holes, but cosmologists eventually built them into effective models of the galaxies. Immense black holes are known to be the powerhouses dwelling at the cores of many galaxies, almost certainly including our own Milky Way. Many giant stars, imploding in supernovae, collapse into spinning black holes.

To begin, let's look briefly in more detail at how the discovery of black holes came about, those ultimate oddities of last century's classical astronomy.

In November, 1915, Albert Einstein caused an unprecedented upheaval in the scientific worldview with his announcement of General Relativity. Basic to his theory, as we've seen, was the Special Relativistic proposal that reality must henceforth be measured according to a new geometry. Space and time, the two basic categories within which we perceive events ("It happened two meters left, three forward, one centimeter up, and ten seconds ago"), dissolved into a single entity, spacetime. This remarkably powerful intellectual leap implied, among other things, that any material body "kinks" or "curves" or even "drags" the spacetime around it.

What's more, matter and energy also melted into a single phenomenon. One of the principal confirmations of Einstein's theory was an observable effect created by the Sun's enormous gravitational field. The effect is piquant. Because energy and mass are identical, the stupendous quantity of energy represented by the Sun's gravity field itself generates a further increment of gravitation, just as the primary field is created by the star's mass.

Newton's laws had long since described the way masses attract each other, but Newton was at a loss to account for the mechanism involved. What Einstein suggested was that masses do not "attract" one another. Rather, they deform or distort the intervening space so that the shortest inertial distance between the two points is a "geodesic" or four-dimensionally curved line.

Within months of Einstein's announcement, the German astronomer Karl Schwarzschild had drawn from it the arresting notion that an object of suitable mass and radius can curve spacetime around it to such an extent that it cuts itself off from the rest of the universe. In mathematical terms, his solution to the field equations involved a singularity; certain values went off the board and became infinite. Usually when singularity infinities turn up in scientific theories it's a strong hint that somewhat is wrong with the equations. Not so this time.

In the vicinity of a Schwarzschild singularity, space is so intensely stressed that you can get in, but you can't get out again. The Sun is about 333,480 times as massive as the earth, but its radius is so enormous that its surface gravity is only 28 times that of our planet. (Not that it actually has a solid surface, of course.) If that great mass could somehow be compressed into a ball six kilometers in diameter, the gravity at its surface would be, in effect, infinitely great. The velocity needed to escape from that surface would exceed the speed of light. So even light itself would not be able to leave the surface of a totally collapsed star, which would have become a black, non-radiant hole in space. (Except for what has come to be called Hawking radiation, a sort of sleight of hand in which one of a pair of particles is trapped and its twin is flung away—a kind of virtual radiation from the hole.)

From a distance, naturally, the gravitational effects on the rest of the universe of such a star would be unchanged. Journalists have a bad habit of describing black holes as if they were ravenous predators, sucking nearby stars into their maws. It's true that something like that happens very close to the event horizon of a hole, but from a distance the star's gravity is no greater than it was before the massive infall. But at its event horizon or imaginary surface, the wavelength of radiation trying to leave would be stretched indefinitely, its photons circling the drain, so to speak, trapping the radiation forever. If it were alone in space, there would be virtually no way to know such an ultimately collapsed star was there. In shrinking down below its Schwarzschild radius, a lone hole would have vanished from the universe.

In the case of a binary star, matters become more interesting, for a black hole can steal the substance of its nearby orbiting companion into an accretion disk of wildly energized infalling matter, emitting gales of X-rays in the process.

At first, this remarkable by-product of Relativity seemed no more than a quirk of the equations. Newtonian mechanics had allowed singularities also, but because Newton was ignorant of the light-speed limit for proper mass such singularities did not entail the disastrous consequences that afflict a collapsed star. Today, with vastly more evidence at our disposal about the nature of the cosmos, the black hole is firmly positioned in the pantheon of stellar wonders—including an immense hole, with the compressed mass of four millions stars like the Sun, near the center of our Milky Wave galaxy.

The first star put forward as an excellent candidate for stellar black hole status was the invisible but instrument-detectable companion of a quite close visible star that behaves so oddly that the presence of a stellar black hole in orbit around it could be inferred. The primary star is HDE 226868, a hot blue B-class sun much larger than our own, some 6200 light years from Earth. The invisible secondary star, Cygnus X-1, orbits it every 5.6 days, and produces intense fluctuating X-ray emissions. Since its mass is estimated to be nearly nine times that of our sun, Cygnus X-1 (which was discovered in 1971) had a strong enough gravitational field to have imploded into a black hole state. In 2011, NASA announced:

Using X-ray data from Chandra, the Rossi X-ray Timing Explorer, and the Advanced Satellite for Cosmology and Astrophysics, scientists were able to determine the spin of Cygnus X-1 with unprecedented accuracy, showing that the black hole is spinning at very close to its maximum rate. Its event horizon—the point of no return for material falling towards a black hole—is spinning around more than 800 times a second.[1]

Amusingly, black hole expert Stephen Hawking lost a 1974 bet with this announcement; he had wagered that Cyg-X1 was *not* a black hole.

Such a collapsed star starts life as a radiant body at least several times larger than the sun. The more massive stars are, the faster and more furiously they burn, and eventually the nuclear flame exhausts its fuel. Without the constant fantastic pressure of stellar radiation blasting out from its interior, the star's monstrous mass collapses under its own weight.

This rapid implosion sometimes provides enough new usable energy to blow away part of the star's substance in a glorious incandescence. Once a century, on average, each galaxy produces an even more dramatic catastrophe: a supernova. When that mighty candle is puffed out, the supernova's core can continue to collapse. The electrons of its atoms are forced into their nuclei, electrical charges equalize, and its entire substance turns into neutrons.

A neutron star is unthinkably dense, yet even this awesome compression is not the limit. If the surface of the neutron supernova remnant sinks within its own Schwarzschild event horizon (beyond which no internal event can ever signal its presence by light radiation) the star…vanishes.

Or not quite. Cygnus X-1, its supernova fires long quenched, is dragging into itself an enormous stream of fiery gas from its highly luminous primary. As this cosmic plunder spirals into the collapsed star it is accelerated to the velocity of light. Gas molecules collide, heating up and shedding gouts of X-rays.

It's important to stress, however, that black holes come in all shapes and sizes, and they are not all as dense as a collapsed star the size of a mountain containing the mass of a star. Seven years after Cygnus X-1 was identified, a group of astronomers reported on an international study of the giant elliptical galaxy Messier 87, which is 54 million light years distant. The team concluded that its core seemed to constitute a black hole which has swallowed up several

---

[1] https://www.nasa.gov/mission_pages/chandra/multimedia/cygnusx1.html

billion stars. This conjecture has been confirmed. M87, a strong source of X-rays, is marked by a jet of hot gas blasting 5000 light years into space from the galaxy's core.[2] Using an excruciatingly sensitive photon-counting device, the astronomers found too few visible stars to account for the mass of the central region, a volume 720 light years across.

It is known that this supermassive black hole nucleus, equal to six and a half billion suns like ours, is surrounded by a rapidly rotating gas ring turbulent with loops and rings created by pressure spasms every few million years. Yet the core's density is not much more than that of air, and if it were not for the colossal bursts of energy raging there it would be quite conceivable to imagine life continuing within its event horizon boundary.

Avid theoreticians swiftly seized on black holes to explain many other outstanding anomalies in astronomy. At one time it was considered possible that dark matter, the 27% of the mass of the cosmos invisible to us, was tied up in that form. (Another vast amount of "antigravity" dark energy, 68% of the cosmos, powers the accelerating expansion of the universe.)[3] Others suggest that the universe in its entirety begins and ends its life by passing through a black hole phase, where all orthodox regularities and physical laws break down totally, perhaps emerging in what is inevitably called a white hole. (None has yet been observed.) Indeed, it's possible that the universe already falls within its own Schwarzschild radius, which means that we would now be living in the heart of a black hole.

The magic of black holes lies in its singularity, the place at its center where physical laws are apparently rendered void. Time and space lose their meaning there. As long ago as 1974, Stephen Hawking gave a lecture at the California Institute of Technology in which he demonstrated that at least one such singularity exists in the universe. Later analysis, much of it by Hawking and his colleagues, cast doubt on that claim, but the possibility remains open. Hawking, as everyone knows, was an authentic genius, tragically pinned to a wheelchair by amyotrophic lateral sclerosis, a progressive disease affecting voluntary muscle movement but not the brain. His mind had wings. He is considered by some physicists to stand beside Newton and Einstein, although even by the time of his death in March, 2018, none of his ideas had met and passed any crucial test.

His early recondite paper was titled "The Breakdown of Physics in the Region of Space-Time Singularities." Wryly, Hawking followed that title with another: "The Breakdown of Physicists in the Region of Space-Time

[2] See https://www.nasa.gov/feature/goddard/2017/messier-87
[3] https://science.nasa.gov/astrophysics/focus-areas/what-is-dark-energy

Singularities." Why? With the breakdown of space and time in their vicinity, the laws of physics become inoperative, or at least our capacity to predict outcomes is lost. Most relevantly to our inquiry, time itself can go into reverse.

Hawking argued that a singularity is not just possible but inevitable when three conditions are met. If General Relativity is true, if gravity is always positive (attracting rather than repelling), and if enough mass has ever been compressed closely enough together, a singularity must exist. Are all three conditions true? As was only learned years later, no. The discovery that the expansion of the cosmos is accelerating rather than slowing, as one might expect, shows that dark energy or some other force or geometrical agency is able to counteract gravity. So conceivably such exotic effects intervene during a supernova collapse—and especially during the extreme conditions thought to have existed during the Big Bang—and prevent the final collapse into oblivion or another universe.

Hawking's most alarming achievement was to weld together the insights of quantum mechanics and Relativity with the proposal, in 1971, of mini-holes. These, unlike collapsed stars, would not collapse under their own weight—they'd have to be squeezed. Unless the material of the Big Bang event was absolutely evenly distributed, which is unlikely, local pockets of immense pressure would have created mini-holes ranging in mass from a planet to a grain of sand. Might the cosmos be choked with a dust of micro-holes? Within 3 years, though, Hawking blew the mini-hole to pieces. If there were ever any mini-holes, they must have gone by now. For black holes aren't black. Taking quantum theory into account alters the original description, permitting holes to lose mass by thermal emission.

This might seem to refute the very postulate that gave rise to the very notion of black holes, but in fact it merely strengthens it. Mini-black holes can evaporate, with an explosion equivalent to 10 million one-megaton nuclear bombs, but collapsed stars and larger black holes still have lifetimes rivaling that of the entire universe. The thermal emission of what we might dub "pink holes" arises because the stupendous gravitational field close to a hole tugs at the very space it occupies, stressing the vacuum. One of the wonders of quantum theory is that empty space is conceived as being full of "virtual" particles, pairs of elementary particles and their antiparticles, that pop into being spontaneously and annihilate one another so rapidly that they cannot be detected. (But their presence can be inferred from certain physical effects on other observable phenomena.) As noted earlier, a black hole can trap one member of such a pair while kicking the other away.

It might be thought that this would add mass to the hole, as well as providing a fuzzy surround of escaping virtual particles. Actually, since the pair-creation occurs at the expense of the hole's gravitational energy, the final upshot is a loss of energy to the hole. In the long term, this is why black holes evaporate.

Interestingly, Feynman's interpretation of positrons crops up again in black hole thermal emission. Hawking declared that we can

> regard the member of the pair of particles that falls into the black hole—the antiparticle, say—as being really a particle that is travelling backward in time. Thus the antiparticle falling into the black hole can be regarded as a particle coming out of the black hole but travelling backward in time. When the particle reaches the point at which the particle-antiparticle pair originally materialized, it is scattered by the gravitational field so that it travels forward in time.

What would it be like inside a collapsed star? One of the advertised features of stellar black holes which endeared them to popularizers is the way they tear incautious explorers to tiny pieces. The gravitational gradient curves so abruptly near a non-rotating collapsed star that an explorer's limbs and organs would be subject to savagely unequal stresses, stretching in one director and squeezing like an Iron Maiden in the other. This is true, however, only in the non-rotating case. If the collapsed star is spinning, an astronaut would be killed only if he or she entered from an equatorial direction, for that is where the infinite tidal forces are located. And it would be quite safe to enter a black hole of galactic magnitude, like the ones thought to lie in the heart of all or most galaxies like our own.

An early student of the lunatic nature of the collapsed star's interior was mathematician John G. Taylor (1931–2012), whose book *Black Holes: The End of the Universe?* appeared in 1973. For a time Professor Taylor became distracted by another phenomenon on the margins of orthodox physics—the metal-bending allegedly performed by psychic and stage performer Uri Geller and others. But while Taylor was clearly a man with a questioning mind, his imagination was constrained by strict mathematical rigor. (He swiftly repudiated metal-bending and telepathy on the conservative grounds that no known force, especially electromagnetism, could produce such effects.)

Once inside the event horizon of a black hole, Taylor argued, we would discover

> a topsy-turvy world indeed, one of the strangest we could ever envisage. For in it, time and space would have been interchanged. In our normal world we can move about freely in space. Time is just the opposite; though we might slow it down near the surface of the event horizon it still marches onwards.

This is exactly reversed inside the event horizon. There we would have no control over our voyage through space, even though we could go backwards in time, or simply move around in it to our heart's desire. (90–91)

It is only fair to point out that not all scientists at the time took the mathematics of black holes and their singularities so literally. Taylor's picture was suggested by strenuous mathematical analysis, but lacking observational confirmation you could take it or leave it. Many physicists were happy to leave it.

There is one ultimate variety of black hole, in all this panoply of paradox, that is more electrifying than the rest. For it's possible that a black hole might form without an event horizon, or be denuded of it through thermal emission.

After all, the surface gravity of a massive body depends on the distance between the surface and the body's center. The faster a body spins, the more it bulges at the equator. The planet Jupiter, for example, is visibly flattened at the poles by its rapid spin. So the influence of gravity on gas giants such as Jupiter is more pronounced at the poles than at the equator. A very rapidly rotating black hole might be flattened into a pancake, so that its event horizon passes within the equatorial zone. Information from its singularity might then spread outward to the universe through the "protruding" section of the hole. Similarly, an evaporating hole might leave its singularity naked for all to observe.

For Taylor, such a possibility was devastating. If we ever locate or create a naked singularity, he warned apocalyptically,

We would be able to travel in time to our heart's desire, to return and meet ourselves, to people our earth with many, many copies of ourself. The time travel which only occurred for the intrepid few who vanished into the no-man's-land inside a spinning black hole would now be available to anyone in the whole of our universe. The trip back in time could only be taken by a suitable voyage round the naked singularity, but it would always be possible. (108)

The seemingly-inevitable causality paradoxes terrified Taylor, but excited many sf writers, as we shall see. What's more, there are analyses that show ways to avoid fatal paradoxes by insisting on certain restrictions. In effect, the blurriness and uncertainties of the atomic world become amplified up to our own scale of existence, saving the time traveller from paradox—but perhaps only at the cost of swapping one potential history for another, consistent with Feyman's picture in which every possible choice is always taken. Careful logical studies show that self-consistency is one key to plausible time-travel (either physical, or via messages to the past), a crucial topic to which we shall return.

One enchanting possibility was suggested by a theorem proved by the notable mathematician Roger Penrose, now an emeritus professor at the Mathematical Institute of Oxford University. This contingency implies that the collapse of a supernova into its own black hole might abruptly open a path to another universe, one we previously had no reason to suppose existed.

These alternative universes are a feature of Superspace, the concept introduced by John Archibald Wheeler. In quantum theory, fundamental particles are not the hard, determinate pellets of matter conceived by nineteenth century atomic theory, but regions where energy interactions can occur with varying probability. The "probability amplitudes" associated with any particle range from the grossly unlikely but still possible (for example, an electron under observation instantly vanishing to the other side of the galaxy) to the highly probable.

Superspace comprises all these probability amplitudes. Taylor noted that

> Near the singularity of a black hole, quantum effects will become important and the plenitude of worlds in superspace should begin to be experienced. But the superspace being explored ever increasingly as the singularity is reached has none of the properties we would expect. The usual ideas of before and after, of what happens 'next', have to be abandoned; even time itself loses any sense, and we must be prepared for the instantaneous jump from one point of space to another if we can once understand how to penetrate and escape from superspace. (103)

Drastic as that sounds, it was penned before Hawking showed that black holes can be sources of radiation. Even without entering the event horizon, we may be able to experience the disturbing revelations of Superspace caused by gravitationally collapsed-star strains upon spacetime. In Hawking's own words: "Indeed, it is possible that the black hole could emit a television set or the works of Proust in 10 leather-bound volumes…" Happily, he added: "But the number of configurations of particles that correspond to these exotic possibilities is vanishingly small."

3

Closed Timelike Loops

In the insane vortex of the black hole interior, then, time's arrow is twisted like a pretzel. If intelligence in the distant future ever learns to ride those currents—perhaps by the deliberate construction of a naked singularity, or a Tipler time gate (see below)—time travel may become commonplace. The ultimate absurdity of science fiction will be a matter of mundane fact.

Our descendants, or their nanoscale AI drones, might venture through history in search of knowledge, aesthetic stimulus, and diversion. A new specialist might be born: the investigative, participatory historian. Temporal anthropologists might study at first hand, or through such miniaturized telemetry equipment, the Paleolithic geniuses painting bison and stags on the cool stone walls of the Lascaux caves, and send their tiny or stealthed recording probes among the aircraft that firebombed Dresden and the billowing smoke of the falling towers of Manhattan.

Alas, we can even imagine the lurid advertisements of hucksters organizing vulgar tours of history's high spots:

"Stroll in PERFECT SAFETY amid the CARNAGE of the BLACK DEATH."
"See the eco-doom COLLAPSE of 2030."

And there are far more bizarre possibilities, all of them explored by clever science fiction stories and novels.

A neo-Nazi revival might travel to the 1930s with the intention of giving Hitler laser or nuclear weapons. Religious time travelers would certainly be tempted to question Gautama Buddha in 528 BC as he sits beneath the bo tree, or revive the crucified Jesus in his tomb, using advanced medical devices.

© Springer Nature Switzerland AG 2019
D. Broderick, *The Time Machine Hypothesis*, Science and Fiction,
https://doi.org/10.1007/978-3-030-16178-1_3

(No doubt some cult has already preached that this is exactly what happened.) On the other hand, it would be enchanting if the reason Mary and Joseph were forced into the stable was the heavy tourist industry pressure from time travelers in Bethlehem that Christmas.

"The most convincing argument against time-travel," Sir Arthur C. Clarke pointed out at the start of the 1960s,

> is the remarkable scarcity of time-travelers. However unpleasant our age may appear to the future, surely one would expect scholars and students to visit us, if such a thing were possible at all. Though. they might try to disguise themselves, accidents would be bound to happen—just as they would if we went back: to Imperial Rome with cameras and tape-recorders concealed under our nylon togas. Time-travelling could never be kept secret for very long… (*Profiles of the Future*)

How would a visiting time traveler appear to us? It is impossible to be sure. Speculation is so fertile when the basic limiting factors are unknown that almost any answer has some degree of plausibility. The best we can do is set reasonable constraints and proceed from there. This is exactly what science fiction writers have been doing ever since H.G. Wells first invented his "chronic argonaut."

The initial set of constraints must concern causality—the web of actions in which one or more events directly give rise to a certain effect, a process which our common experience assures us can take place only in a single, forward direction of time. You can beat the egg into an omelet, but you can't unbeat your savory steaming breakfast back into a shell-sealed egg.

Most philosophers have rejected time travel as a self-contradiction (as they have also rejected psychic precognition of the future, despite laboratory evidence in its favor), because the paradoxes entailed can oppress the mind with panic.

The hoariest chestnut runs as follows: What happens if you build a time machine, travel back to a point before the conception of one of your parents, accidentally or deliberately kill your future grandfather, and hence short-circuit your own existence? If you have never existed, your ancestor cannot have been killed by you, so you will be born after all, only to kill him, in an infinite vicious loop.

The hokey solution is to invoke some authority similar to the Coast Guard. Call them the Time Police or Time Patrol. If a temporal paradox occurs, an alert squad rushes back through time to the scene of the initial action and nips it in the bud. The reason we never observe time line changers, then, is that they are all in prison, apprehended before their crimes were committed. It raises some interesting jurisprudential issues. The movie *Minority Report* played with a version of this notion, based on a story by Philip K. Dick, a brilliant science fiction writer who tried out just about all the variants on time loops and precognition.

But the logistics of that solution, appealing though it is to writers of adventure stories, are insupportable. At the very least, the conjecture tends to catapult itself immediately into a scenario of grotesque wars between various epochs, each striving to undo and remake history to their own taste (a premise exploited in a long series of stories about the Time Patrol by the equally brilliant Poul Anderson).

Even if this is the case, it would be to the advantage of time travelers to avoid paradoxes. If visitors from the future are thronging our streets, they will take care not to reveal themselves in any circumstance where such intervention would damage their own history. When a time machine gets into trouble, we can expect that maintenance backup crew will shuttle back and pull it out before someone in the innocent past stumbles over it.

Time machines, if we did observe them, might well have a tendency to appear and disappear abruptly, unless they must first voyage through space to the vicinity of a naked singularity. To a naive observer, this would look like "dematerialization" or "teleportation."

A single temporal device studying a given vicinity could flip from date to date without changing its geographical position (assuming that it could compensate for the motion of the earth through space, an engineering problem facing all time machine designers). The impression would thus be created of a space vehicle returning to the spot at different intervals. For the time travelers, the experienced sequence would be like a time-lapse film run straight through, while for conventional observers there might be a gap of weeks or years between visits.

If we were very lucky, one way to distinguish such time machines from space vehicles could be a periodicity in the visits relating to our arbitrary calendars, rather than to the time it would take to get here from an extraterrestrial origin. In other words, if anomalous craft were seen dropping in on the first of every month, we might suppose that their occupants employ the same arbitrary calendar as western twenty-first century cultures, or at least know about them.

Even more radically: if time travel of a psychic kind is ever developed (perhaps a machine-boosted form of remote viewing), selective telepathic contact with the past might prove feasible. A future historian might tune his amplified psychic presence to an alchemist in the Middle Ages, appearing to him as a "demon" whom others would be incapable of seeing.

It is notable that throughout recorded history, in many dispersed societies, momentous events are said to have been attended by "comets," "signs in the sky" not always attributable to stellar or meteorological conditions. What's more, astute if superstitious soothsayers have claimed that these "omens" precede the events, as well as accompanying them. It is as if someone interested in the crisis had known it was due, and had checked out the circumstances leading up to it.

Astronomer Carl Sagan, an early proponent of life beyond the Earth, estimated that our planet can expect a visit from extraterrestrial sources on average only once every thousand years. But if such craft possess temporal maneuverability to match their interstellar capacity (and the latter, providing access to black holes, might permit the former), we could expect observation from alien cultures that discover Earth not today but in millennia to come.

A truly gruesome explanation for their interest in our present era cannot be discounted. In the five billion years remaining before earth is crisped in the Sun's terminal flame, many thousands or millions of alien spacecraft might stumble on a wrecked, desolate planet, and wonder at the poignant ruin they behold. Filled with curiosity and pity, they might follow back the four-dimensional worldline to the bitter epoch leading to the time when human civilization finally destroyed itself and its world: the second half of the twenty-first century…

In his excellent book *Time Travel in Einstein's Universe*, Professor J. Richard Gott explains his motivation in pursuing this apparently loony topic:

> Why are physicists like me interested in time travel? It's not because we are hoping to patent a time machine in the near future. Rather, it's because we want to test the boundaries of the laws of physics. The paradoxes associated with time travel pose a challenge… a clue that some interesting physics is waiting to be discovered. (29)

Self-consistency is one way such researchers have found to evade what seem to be an assured deadly outcome. It allows visits to the past only if no change is made to a history known to have occurred in a certain way. Gott observes that

> arriving at self-consistent solutions—in fact, numerous ones—always seems possible from a given set of starting conditions, as suggested by Thorne, Novikov, and their collaborators in an elaborate series of thought experiments involving billiard balls going back in time. They tried to produced situations where a time-

traveling billiard ball would collide with its earlier self, deflecting its trajectory so it couldn't enter the time machine in the first place. But they could always find a self-consistent solution where the collision was only a light tap that didn't stop the ball from entering the time machine, but sent it on a path that made it nearly miss its earlier self... (20)

It's necessary to look briefly at the way physics currently describes the fundamental stuff of which the universe is composed, whether seen as particles or fields. For it is in the realm of the ultra-small that we witness the boundary between ordinary reality and the probability realm of Superspace.

High-energy physics is the science concerned with the ultimate constituents of matter, and after many years of creative confusion it has seemed for several decades on the verge of a powerful synthesis. Even so, the state of play in both theory and experimental findings alters almost weekly. The results can be surprising and ironic. In 1881 Albert Michelson first found evidence that the supposed universe-filling aether, through which light and other waves were supposed to swim, did not exist. Now, thanks to Large Hadron Collider experiments, an equally pervasive and mysterious phenomenon, the Higgs field (manifested as an identifiable boson particle), *was* shown in 2012 to exist. Physicist Peter Higgs and a colleague shared a Nobel prize for this discovery in 2013. The Higgs field, which gives other particles their mass, looks oddly like a tuned-up version of the nineteenth century aether.

The old picture of the atom, composed of a nucleus of protons and neutrons orbited by planetary electrons, has long since broken down, though it remains a useful simplification in dealing with certain chemical reactions. Enormously expensive experiments revealed a bewildering array of more than 300 "elementary" particles, most of them so-called "resonances" so unstable that they exist for only fractions of a millionth of a second before decaying into new forms.

Each is characterized by specific values and properties, some well-known—mass, charge, spin—others arcane—strangeness, charm, and so on. These create binding and repelling forces by shuttling quanta (or "energy packets") back and forth between them. And, crucially, not all the particles are subject to all the interactions, or forces.

To date, science knows only four kinds of basic forces (or maybe three), though several candidates for a fifth have been proposed. Until some 40 years ago, it was thought that none of them was reducible to any other. They are gravity (the feeblest); the weak nuclear force (10^{25} times stronger than gravity) and electromagnetism (10^{36} times), which are really manifestations of an electroweak force; and the strong nuclear force (10^{38} times stronger). That is, the strong nuclear force is more powerful than gravity by a factor written out fully as a one followed by 38 zeroes. But the strong force is correspondingly limited in range, operating over a maximum distance of a trillionth of a millimeter.

The concept of "force" itself is frequently seen as only a convenient abstraction for the varieties of interaction between elementary particles. These interactions are thus the result of exchanges of energy bundles or quanta. For example, two electrons repel one another electrically by swapping a photon, the radiant particle that, at different frequencies, constitutes gamma rays, visible light, heat and radio waves. Nucleons such as the proton and neutron (themselves comprised of more fundamental particles) are bound by an exchange of pions (giving rise to the strong force) and the W or intermediate vector bosons (which carry the weak force responsible for radioactivity).

The story is far more complicated than that, but paradoxically it may also be simpler as well. In a sense, all the particles that engage in strong nuclear interactions (collectively termed hadrons, and including the proton and neutron) represent various excited states of just three point-like particles. It was realized half a century ago that the hadron resonances could be gathered into elegant groups according to a symmetry dictum, and this was confirmed experimentally three years later. It was quickly surmised that this regularity was due to the hadrons' further internal structure, that they were composed of new, massive particles. This new class of fractionally-charged particles was dubbed *quarks* (rhymes with "corks") by Murray Gell-Man, subsequently a Nobel laureate in physics. These quarks, with fractional charge, are stuck together, inevitably, by gluons—which come in eight varieties, color-coded. The force holding individual quarks together gets *stronger* as they are pulled apart, so they have never been seen.

Particles called leptons, involved in weak nuclear interactions—the electron, the muon or "heavy electron," and the tau, the heaviest, and their separate neutrinos—do seem to be genuinely elementary. At the same time, by a series of stunningly bold intellectual gambits, it was shown that a theory termed "broken gauge symmetry" can unify the weak and the electromagnetic forces into a single electroweak force, and the ultimate goal—still not realized—is the compression of all four known forces, including gravitation, into a single force broken asunder near the Big Bang, as the expanding cosmos cooled, by the breaking of that original symmetric unity.

There is something magnificent in a physics where, in the words of Dr. Christopher Llewellyn-Smith, "It turns out that although different parts of a calculation seem to yield meaningless results, the underlying symmetries lead to a wonderful cancellation of all absurdities leaving a sensible result when everything is put together at the end."[1] The quantum theory explanation for the time traveler's Grandfather paradox could be evaluated in the same words. But we have not quite finished laying the groundwork for that explanation.

Simplicity in one place demands complexity in another. In short, perhaps, to borrow the motto of the United States: *E pluribus unum*. Baryons are thought to be composed of three quarks, while mesons are built from a quark and an antiquark. Gauge theory ended by requiring another three quarks (which ruins the Joycean poignancy of the name—"Three quarks for Muster Mark," wrote James Joyce in *Finnegans Wake*), but apparently no more than that. The glorious metaphorical properties color, charm and flavor joined charge, isospin and strangeness. (Who said there was no romance, or at least whimsy, in science?)

Straining our minds to the limit, we must understand that all these particles that together make up the matter around us interact dynamically with the vacuum of space. The zero-energy state of raw space seethes with activity we cannot detect directly. It is the source, you'll recall, of the "virtual particles" brought into being by the intense gravitational field of a black hole. These pseudoparticles possess no direct physical interpretation, but they are not to be scoffed at.

Consider a discussion by Dr. Sidney Drell (1926–2016), a professor emeritus at the Stanford Linear Accelerator Center, an accomplished violinist, a winner of the 2000 Enrico Fermi Award and a recipient of the National Medal of Science in 2012. The foundations of modern physics, he once observed, are the conservation laws, of which the most inviolable are those relating to energy and momentum. These quantities can be transferred, but "the sum of the energies and momenta before an interaction must be exactly equal to the sum afterward."

Alas, when an electron and positron meet headlong in a storage ring accelerator, each bearing the enormous mass/energy the device has imparted, and annihilate into pure light, the event breaches the most hallowed of conservation laws. Do physicists panic? No.

For the emitted photon, which decays instantly into a range of other particles—re-creating mass from energy—is not a "real" photon. "It cannot be real," Drell stated, "because it has the wrong proportions of energy and momentum."

Hence it is dubbed a "virtual photon," and "its most important characteristic is that it can never be observed... because it decays before it can be detected. According to the uncertainty principle formulated by Werner

---

[1] *New Scientist*, Apr 10, 1975, 77

Heisenberg, the lifetime of a virtual particle is necessarily too brief for the particle to be observed."

This is just as well, for if it could be observed it would be seen to be an impossible entity and it would be obliged never to have existed. So to speak.

Nor is this almost incomprehensible effect restricted to black hole regions and linear accelerators. Stephen Hawking once noted:

> Quantum mechanics implies that the whole of space is filled with pairs of "virtual" particles and antiparticles that are constantly materializing in pairs, separating and then coming together again and annihilating each other ... They cannot be observed directly with a particle detector. Their indirect effects can nonetheless be measured, and their existence has been confirmed by a small shift (the "Lamb shift") they produce in the spectrum of light from excited hydrogen atoms.[2]

The physics of rotating black holes and naked singularities was first broached by the astronomer Roy P. Kerr in 1963. It is the "Kerr Metric" that leads to the portrait of wormhole spacetime channels hurling material instantly across a billion light years, with a corresponding shift backwards through negative time to a billion years in the past. But a decade later, a still more striking conception was announced by Frank J. Tipler (b. 1947), in *Physical Review D*. A mathematical physicist, Tipler still holds joint positions in the Departments of Mathematics and Physics at Tulane University. He argued that the principles of General Relativity, coupled with the properties of a particular kind of extremal phenomena somewhat similar to the black hole, allowed the design of a two-way time machine.

The late Dr. Robert L. Forward, an expert researcher on gravity at the Hughes Research Laboratories, summarized these findings this way in his popular science book *Future Magic* (1988):

> Tipler's time machine is a long cylinder of ultradense mass with a spin speed at its surface that is one-half that of light. The time-mixing region is near the midpoint of the cylinder, but *outside* the mass... The important feature of the Tipler Two-Way Time Machine is that it allows travel both backward and forward in time (depending upon whether you circle with or against the spin of the cylinder) and neither the time traveler not the time machine has to move at velocities close to that of light. (175)

---

[2] Stephen Hawking, 2nd Reith Lecture, 2016: https://www.bbc.com/news/science-environment-35421439

This, you'll note, is a great convenience for time machine operators, and a considerable advance over diving through the murderous event horizon of a collapsed star with a pretty definite guarantee of no-exit. Cosmic strings created at the Big Bang could fit the bill. Tipler concluded that an infinitely long cosmic string might not be needed, replaced by this constructed rotating cylinder perhaps a hundred miles long and containing the compressed mass of several black holes, spinning at half the speed of light. This, he argued, could create closed timelike lines that would permit time travel without necessarily ruffling the passenger's hair. A few years later, Tipler found that special "exotic matter" would be needed, and even that might not be sufficient.

Tipler's approach was, at the time, a brand new idea to physics, and was quickly adopted in science fiction as "T machines" (manufactured bequests of an ancient interstellar civilization) by Poul Anderson in his novel *The Avatar* (1978).

A massive cylinder of the kind required to meet Tipler's specifications probably does not exist in nature. A quite different method (in principle) for traveling into the past and then back again has been explored at length by Richard Gott. It depends on the possible existence of cosmic strings, monstrous residues of the Big Bang epoch, where colossal amounts of energy are trapped inside fantastically narrow "strings" that might stretch across the entire universe, and under vast tension. By flying *around* such a string in a certain trajectory—assuming you could find one and reach it—you would be flung into a different time. Better still, corral two of these strings and juggle them until they are stretched parallel to each other. When people talk about "playing god," this is the sort of scale that seems appropriate, so it will require an immensely powerful supercivilization to manage this gateway. Worse, as Gott points out in *Time Travel in Einstein's Universe*:

> A collapsing loop of string large enough to allow you to circle it once and go back in time one year would have more than half the mass-energy of an entire galaxy. But a worse problem exists—such a massive string would become so compact as it collapses that it would be in danger of forming a black hole. (110)

And nobody wants to get caught by one of those.

It is not impossible, though, that sometime in the future our descendants will have the prowess and energy resources to construct one. If that day comes, time travel will be a reality.

If so, isn't it plausible that those descendants will return through time (or send probes high in the sky, or perhaps invisibly small) to view us, their ancestors? It is equally possible that if intelligence is arisen on other worlds, already in existence or evolving in the future, alien beings might search not only the vast expanses of our current universe but also the deep multi-billion year epochs of past and future. And find us, here and now.

Another dramatic novelty of the second half of the last century was the possibility of tachyons, particles that travel *only* faster than light. Unlike Feynman's antiparticles, whose proposed motion backwards in time is a matter of interpretation, tachyons must exist in a condition of time reversal (if they exist at all, and so far there is no evidence for that).

Strangely enough, these hypothetical particles do not abrogate Einstein's dictum that nothing can accelerate through and past the speed of light in vacuum, for tachyons *always* have velocities swifter than light and cannot go slower than that. They arise, in fact, as a direct consequence of Relativity. Since they are a kind of reversal of our sub-luminal physics, they possess certain bizarre properties. The *slowest* state of a tachyon is just barely above the speed of light, and it spends most of its existence there. As it loses energy (from a trail of gravitational radiation, Gott argues), its momentum increases, until it is traveling infinitely fast—its worldline smeared out across all of the universe. Why and how? In his 1992 book *In Search of the Edge of Time*, John Gribbin noted that the speed of light in a medium such as water is lower than the vacuum velocity specified as the faster speed possible. So a charged particle moving faster than the ambient limit sheds energy in a flash of photons, an effect named for its discovered, Pavel Cherenkov. Hence,

> A charged tachyon, moving faster than light even in a vacuum, would also have to emit Cherenkov radiation, as long as it has any energy available to radiate… [losing] all its energy literally in a flash, ending up with zero energy and traveling at infinite speed. (210)

So even if we had a way to locate tachyons and couple them to our instruments (and would-be time machines), they would be difficult if not impossible to steer. Their importance to the Time Machine Hypothesis, if they prove to be real, might become paramount as a means of communication. In Gregory Benford's notable novel *Timescape*, they become the means of transmitting an urgent warning message to the past and thereby saving the world (or *a* world) from a planet-damaging blight. Unfortunately, if Gott is correct, "tachyons would spend most of the time moving at just barely over light-speed. Therefore, tachyons could not be used to send energy or information faster than light over macroscopic distances" (Gott 2001, 127–28).

A far less extravagant project for reclaiming the past has been the main undertaking of Professor Ronald Mallett, an African-American physicist whose father died at 33 from a heart attack induced by heavy smoking, when Ron was just 10 years old. Reading a comic book version of Wells's *The Time Machine* fired the child's imagination and set him on the path to devising a time machine of his own, hoping to travel back to his childhood and persuading his beloved gadgeteer father to abandon smoking. In a 2018 BBC Horizon science program titled *How to Build a Time Machine,* he declared: "I think of myself as being an ordinary person with a passion, and my passion is the possibility of time travel."[3]

This passion led him to computer work for the US air force and then a General Relativity PhD from Penn State University on time reversal in a de Sitter universe, one of the early cosmological models that had predicted acceleration of the expanding cosmos only proved in 1998. He knew that massive objects subject spacetime in their vicinity to the phenomenon of "frame dragging," where the fabric of space and time is twisted by powerful gravity—something predicted especially of stellar-scale black holes, but caused even by the Earth's gravity.

His hope, expressed in an experimental laboratory research program, is that frame dragging on a small but detectable scale can be initiated and sustained in a ring laser. The gravitational field of intensely energetic light beams would twist local spacetime into a kind of helix that generates a closed timelike loop, where the temporal direction is reversed. He published his findings in 2000. So far he has only a tabletop device, potentially capable of firing particles into the past, showing proof of concept and eventually thus transmitting messages but nothing on the scale of a human being. That would be a remarkable and literally history-changing invention, but it would not be able to save Professor Mallett's father's life because theory insists that time machines would be limited to the period following their earliest activation.

Since 2013 to the present, he has been Professor Emeritus and Research Professor, Department of Physics, University of Connecticut. Performing such experiments on hand-built rigs is necessarily expensive, and getting such a device to the stage of practical time reversal would cost millions of dollars. As I write, Mallett is already 73 years old, and his opportunities to gain grants for such work are limited. The director Spike Lee holds the rights to make a feature-length movie of his quest, and if that eventuates perhaps sponsors will fund further research.

---

[3] https://www.bbc.com/news/science-environment-44771942. Mallett's visionary quest is discussed in moderate detail in a chapter (228–36) of Brian Clegg's book that shares the same title with the BBC program and Paul Davies' book, and in a Wikipedia entry at https://en.wikipedia.org/wiki/Ronald_Mallett

Meanwhile, though, skeptical academics have found fault with his analysis. Several serious objections were raised by Ken D. Olum and Allen Everett in 2005,[4] starting with the exceptional energies need to drive it, and a calculation that the ring of lasers would need to be larger than the circumference of the universe. At one time Mallet made what seems an elementary error in proposing that slowing the laser light via a superfluid medium would solve some of the problems, but as Richard Gott argued, this confuses such confined effects with the speed of light in a vacuum (c) which is the relevant parameter in General Relativity theory. It seems unlikely that Ronald Mallett's search will lead us to a pathway into the past.

---

[4] "Can a circulating light beam produce a time machine?" (*Foundations of Physics Letters*. 18 (4): 379–85).

4

Don't Change the Past

The most lucid book on black holes and time warps, published by Professor Kip Thorne in 1994, is titled, unsurprisingly, *Black Holes and Time Warps*, and remains both enjoyable and authoritative more than 20 years later. Startlingly, perhaps, the final chapter is titled "Wormholes and Time Machines," and Thorne is careful to provide a footnote taking responsibility for writing the book "solely from my own personal viewpoint" (483). It is a rich resource on the foundations of both topics, and has some detailed and clearly illustrated accounts of possible time machines.

This topic was horrifying to the era's solemn custodians of science, so Thorne and his friends were initially cautious lest they be represented as lunatics or "sci fi" cultists. It was a real concern. In 1987, when Thorne was first convinced that wormholes might become time machines (of a kind), a close friend told Thorne's wife of his concern that Kip had "gone a little crazy or senile or…" Thorne realized that care would need to be taken "about how we presented our research to the community of physicists and to the general public." When these findings were published at last, the anxious friend remained anxious that reputations would be sullied. Luckily, the research program was not slain in the crib. Thorne notes:

> My Russian friend Igor Novikov, by contrast, was ecstatic… "I'm so happy, Kip! You have broken the barrier. If *you* can publish research on time machines, then so can *I*!" And he proceeded to do so, forthwith. (505, 508)

It is fascinating to track the papers they succeeded in placing in *Physical Review D* and other heavy-duty journals, starting more than 20 years ago.

© Springer Nature Switzerland AG 2019
D. Broderick, *The Time Machine Hypothesis*, Science and Fiction,
https://doi.org/10.1007/978-3-030-16178-1_4

These new analyses were compelling—no doubt because the power of general relativity was employed as their basis—and the documents are now part of the canon, even if still at the boundary. We can read explorations of closed time-like curves and wormhole time machines by Kip Thorne and associated scholars: Joe Polchinski (1954–2018), Fernando Echeverria (whose research with Thorne included "Self-consistency in wormhole spacetimes with closed time-like curves"), Gunnar Klinkhammer (whose 1992 Caltech dissertation was on Multiply Connected Spacetimes and Closed Timelike Curves in Semiclassical Gravity), Ulvi Yurtsever (work on Classical and quantum instability of compact Cauchy horizons in two dimensions), Richard Gott, and others.

One of the classic arguments attempting to refute the possibility of time travel to the past is the apparent inevitability of messing with the present and future by making changes to past events. This, it was felt, had a high probability (and paradox-creating risk, as noted above) of preventing the traveler's own existence. A disturbingly amusing sideways glance at such effects is provided by a Canadian cartoonist who uses the 'nym "Winston Rountree."[1] In "Doubt creeps in," two early Nazis guard Chancellor Hitler's 1933 quarters. Jürgen comments in frustration, "I know that Herr Hitler is a great leader… it's just…" With a sizzling *Brrrrzt* a futuristic warrior burst out of nowhere into the corridor. Jürgen instantly shoots him dead, and finishes, "…it's these constant attempts on his life by time travelers. I mean, you kind of have to wonder." In an episode of *Doctor Who*, "Let's kill Hitler," an attempt on Hitler's life by a companion of the time traveling Doctor actually rescues the Nazi leader from imminent accidental death.

This kind of extended temporal causality can be generalized: most people alive in the era of time travel will have ancestors whose lives would be thrown off course if dreadful tragedies in the past were prevented retroactively from happening. My Australian maternal grandfather fought Germans in Europe in the blood-drenched catastrophe of the "Great War." Wounded, he ended up in England and met his future wife, who would become my mother's mother. If World War One had been averted somehow, I would not be here to write these words—and perhaps you would not be alive to read them.

Why? Because, even with the most *trivial* interruptions to an event on that scale, almost all humans alive now would not have been conceived in the resulting "modified history." Paternal spermatozoa swarm in the testes like random jittery particles revealed in a beam of afternoon sunlight, batted back and forth in Brownian motion. With global war prevented, most of those soldiers and non-combatants who would have been killed remain alive, many

---

[1] http://www.viruscomix.com/page382.html

become parents, their presence in the world necessitates ever-widening impacts small and large. Perhaps this would have little impact on people in regions of the planet remote from that particular war, but World War II might have come earlier (or been averted, who can say?), and in either case people's lives would have been jolted in various degrees.

This surge of tiny changes in activities, locale, technology, even clothing styles would have rippled across the planet. In surprising consequence, the probability of a certain spermatozoon entering the same ovum needed to create *you* is infinitesimally small, because there is no selection process driving the timing or outcome.[2] In miniature, this same insight is at the poignant heart of the 2013 movie *About Time*, where a much-loved girl infant is replaced by a boy because a crisis had caused the father to change the timing of his love-making.

You can't go home again (to the womb), because you were never there, this time around. That's Ray Bradbury's famous butterfly effect (carelessly kill a butterfly in the deep past, as his story "A Sound of Thunder" proposed, and everything in your own history will change) writ incomprehensibly large—and self-refutingly.

One of the most amusing mass media examples of this hazard is the movie trilogy *Back to the Future*. Marty McFly's appearance in 1955 almost prevents his parents' marriage and his own birth due to his future mother's sudden infatuation for her future son—before any of these events have even happened. (A studio executive who saw the script offered by writer Bob Gale and director Robert Zemeckis blurted "We can't make this movie. Are you guys *insane?*" A light-hearted teen movie about potential *incest?* But it was made, and is now a beloved classic.)

Marty's blundering keeps changing small aspects of the future, which causes ripple effects slowly percolating to and from the altered future. This is represented, absurdly but effectively for a non-scientific mass audience, by photos or even body parts slowly fading out of existence until a wild stab at a remedy prevents or skews the disaster. (The McFly domicile is set in an enclave near Twin Pines Mall; after one temporal debacle, its sign now reads "Lone Pine Mall"…) Doc sketches on a blackboard a track of events that branches at some point, creating another universe with a slightly variant history. The rest of the trilogy toys with this notion of alternative histories, with numerous Martys appearing from alternative futures, mostly played for laughs and

---

[2] Well, unless you believe a god or some equivalent supernaturally powerful entity carefully brought about exactly the same happy event, including all those many cases where the resulting baby was born with injuries or deficits that can't be repaired.

thrills. Some theorists, including the British-Israeli quantum specialist David Deutsch (b. 1953), do propose something like this many worlds model as a solution to the paradox of changing your own history, even when that should obliterate you. Instead, you can continue on, but in a separate universe.[3]

Novikov found a kind of solution in his Self-Consistency Principle which simply banned paradoxes, causing them to fizzle. You can go back in time except where what you do is flagrantly inconsistent with established history, which then short-circuits or resets the process. Why and how? Well, things are just that way. You can't walk upside down on the ceiling, either, Novikov reminds us. This strikes me as a rather feeble word game, but it led to some extremely interesting further thinking by Thorne and his colleagues.

Let's go back to the wormholes, following suitable preliminaries (first get your wormhole, perhaps from the very quantum dust of reality in extremely pre-expanded form, and inflate or preserve it with suitable shaped exotic matter, if you can find any—alas, none has yet been detected). Now open one end of the hole in your lab, and dispatch the other at speeds approaching that of light. Alternatively, put the distant end in orbit around an immensely massive condensed object such as a black hole. Either method will retard the length of time experienced by the traveling worm-porthole. Return it to Earth a century later (by reference to the Earth's spacetime frame), and the view looking into the distant wormhole mouth will show a past still barely changed since the craft achieved near-light speed, or black hole near-orbit.

This is just the so-called "twin paradox" of relativity, which notes that a twin traveling near light-speed to a distant star and then returning to Earth at the same speed will find that a much briefer time has elapsed for her than experienced by the stay-at-home on Earth—who indeed might be dead of old age by then. But since this effect is claimed by Einstein to be *relative*, why not choose to see this as the twin on Earth receding and then approaching the starship? No, and this isn't a real paradox at all—just a consequence of special relativity's spacetime shenanigans, which depend on the *changes of acceleration* on the traveling twin—something not experienced by the sedate Earth sister.

Joe Polchinski came up with a neat thought experiment, described briefly earlier by Gott, to test Novikov's principle that requires *consistency*. He uses a small wormhole double time gate, a mechanical arrangement in which a billiard ball is fired toward the wormhole's future-ward mouth, which is coupled backward-in-time to the present-time end. What happens if the ball, as it

---

[3] For a handy summary of the many timelines in *Back To The Future*, consult https://scifi.stackexchange.com/questions/20004/was-there-really-a-1955-enchantment-under-the-sea-dance-timeline-with-just-one-marty

passes the present mouth, is struck by its own future self emerging from the present-time hole even before it can reach the future-ward end and get directed back to the present? If this happens, the infinite cycle that would exist, with the ball going from future to past, back to future, back to past, etc., would never occur. Paradox! Well, not necessarily. It turns out that there is an alternative possible outcome. Perhaps the later versions of the billiard ball strikes its earlier self only a glancing blow, so that the original state of the ball is only slightly diverted from its planned path. Then it does enter the hole and get sent back to its past (our present), and again and again and again, so consistency is sustained.

I prefer to recast this paradox (purely imaginary; no pooch, however ferocious, will be harmed) as a "pair o' dogs," which actually has only one dog, two at a time. Gazooks, a fearsome, loudly barking pit bull, races toward Mouth 1 as she has been trained to do, to be rewarded with a fine piece of steak. As she passes Mouth 2 on her way to Mouth 1, a fearsome and loudly barking pit bull barrels into her path. The dogs thud into each other, turn, pass, pause, noses twitching, and express their delight at this happy meeting. The pair o' dogs cavort, but eventually one or the other remembers the promise of chow and dashes forward to Mouth 1, and a nice plate of steak.

The critical point here is that the adventure with Gazooks was *not* a paradox. It is only our temporal parochialism that makes it seem unbelievable and maybe even immoral that what is plainly a meeting between two very similar canines is actually just one dog interacting with itself, each marked by a stopwatch time showing which is earlier and which later. Her fraught meeting with herself is precisely not a case of self-inconsistency. That would be this impossibility: Gazooks runs toward Mouth 1. Before she can get close enough to step across the threshold into the time bubble and then exit Mouth 2, her doppelganger erupts slavering from Mouth 2, seizes her by the scruff of the neck, and they race away together, snapping and snarling and enjoying themselves tremendously. Neither of these instantiations returns to go through Mouth 1, which means that none of this happened. How could it? Where did the second version of Gazooks come from? She didn't; they never see each other again, because there was only one of them, the impossible dog, never duplicated. Consistency Fail!

There are no paradoxes in time travel. All that can ever happen is that before you enter the past, everything you are going do there, whether you plan it or not, *has already happened* and is part of your history. To think you have any capacity to change what has already happened (even though you're doing it for the first time) is as foolish and illogical as supposing that, like a punk

Alice, you can look into a mirror and see your reflection turn away from you and then spin back, with a bright red painted nose.

But maybe there is always another way for Gazooks to bump herself without inevitably deflecting her path away from Mouth 1 and thereby zeroing out the entire event into ontological nothingness. It turns out, as Kip Thorne tells us, that a lot of further work by him and his colleagues showed that there is not only one way to keep the cycle from obliterating itself. There are, it turns out, a hundred, a million, an infinite number of possible paths that allow a Self-Consistent outcome. Assuming, of course that wormhole time-passage to the past, then back to the present, really is possible.

Stephen Hawking claimed it wasn't, and cooked up his Chronology Protection Conjecture. Spacetime just won't allow it, he said confidently—and not just in the special cases of Gazooks and the billiard balls. Even with exotic matter bracing both doors, a wormhole inevitably (he thought) falls prey almost instantly to a ferocious internal attack from the shrieking, ever-multiplying, reduplicating swarms of the same environmental or even virtual particles being flung again and again from Mouth 1 to Mouth 2 and out and back, redoubling in energy with each circuit, until auto-demolition explodes the time machine to shreds and stray quarks.

So by setting up an impossible causality loop, we return to the apparent paradox of a time traveler aborting his or her own existence. A science of physics based on quantum theory suggests at least one answer flapping away from the problem like a pair o' ducks, a way for the loop-entangled traveler to escape the paradox cycle by shifting into some innocuous sidetrack, one of the alternative histories required by Feynman's "sum-over-histories" model of quantum action. Here is, maybe, another and quite practical answer to the Grandfather and related Paradoxes:

Any serious time machine will need a portable power source. (Although it's amusing to imagine a time machine tethered to an indefinitely long power cord stretching from its initial date to wherever in time it ends up, past or future.) Suppose a traveler makes a choice that constitutes a breach in his or her own causality or unbroken spacetime world line. Perhaps any impossible, self-annihilating causal loop trips a kind of circuit breaker in reality, throwing the machine and its passenger back to the moment of departure. Would memory be retained from this slice of history that never actually happened? I suspect it wouldn't.

In any event, a certain amount of energy will have been expended to make this pointless, self-defeating trip. We can imagine, with a frisson of horror, a long, very long sequence of departure-and-return trips, each draining a small amount of power from the machine's reserves. Let us hope the traveler does *not* remember these cumulative recurrences. But at length, we might suppose, the battery or power generator is exhausted—so the time machine simply refuses to budge next time an attempt is made to begin the fraught trip into the past. The benefit of this notion is that no hand of a deity, nor cosmic law protecting chronology, is required to explain the inevitable escape from the causal loop.

On a much grander scale than a single hapless chrononaut trapped in a loop, a related mechanism has been conjectured by cosmologist Paul Davies. Instead of a universe that oscillates through an infinite sequence of black hole deaths and white hole Big Bangs, Davies suggested that time reverses direction at the end of each cycle. If so, the universe is something like a movie played to the end and then run backwards, only to be shown again, forever. But is the cycling film analogy exact? Must such a universe literally repeat every event into infinity? Dr. John Gribbin has proposed a solution akin to my own wild guess about time travel loops:

> Every time some physical system is confronted with a "choice" among possible quantum paths, it is believed to choose among the possible outcomes at random... Suppose there is one Universe. That there are no parallels, but that the *same* Universe is repeatedly offered a choice between the same huge number of quantum paths. *There is no reason to expect it to choose the same quantum alternatives from those allowed every time the choice is offered.*[4] [Gribbin's italics]

I do not find it totally unlikely that similar considerations might apply, on an ultra-miniature scale, to a time traveler trapped in a potentially self-negating loop.

Was Thorne convinced by such arguments? Not entirely. Indeed, a couple of decades later, he became the science advisor for a highly expensive movie that took wormholes and time contortions very seriously.

Black Holes and Time Warps is therefore a useful companion to Thorne's recent and up-to-date *The Science of* Interstellar, which is both an explanation of what exactly is going on in the puzzling mega-movie *Interstellar* (tenth highest grossing film of 2014) and a vivid illustrated journey into the mysteries of M-theory, the current best attempt to account for everything in our universe and beyond, using string theory. Time travel, of two kinds, is the heart of the movie's plot.

In 1984, this specialist in gravitational theory came to the notice of science fiction fans and scientists alike by guiding the astronomer Carl Sagan away

---

[4] John Gribbin, *White Holes* (1977, 169–70).

from a black hole stargate to a more plausible wormhole, in Sagan's sf novel, and then movie adaptation, *Contact* (1985). This was no back-of-the-envelope advice; Thorne worked hard to calculate the effects required by a supercivilization to get a human from Earth to a distant world in, effectively, nothing-flat, and then home again. His subsequent involvement with the Christopher Nolan-directed *Interstellar* was even more taxing. Begun with a film script he wrote with producer (and, briefly, girlfriend) Lynda Obst in 2005 (*The Science of* Interstellar, 1), they aimed at "a blockbuster movie *grounded from the outset in real science*… A film that gives the audience a taste of the wondrous things that the laws of physics can and might create in our universe, and the great things humans can achieve by mastering the physical laws" (2–3). It was an ambition reminiscent of the drive for scientific and technical excellence behind Kubrick and Clarke's *2001: A Space Odyssey*, but with more heart and passion.

This early script was commandeered by Chris Nolan and his brother Jonathan, but their additions and alterations were intelligent and effective in conveying Thorne's vision into media terms that avoided as much explicit exposition as possible. That made for more special effects-driven box-office success, but left plenty of its viewers groping literally in the dark for understanding. It's simply not feasible to cram a doctoral degree's worth of high-energy physics and general relativity into several hours of space adventuring. Which provided the incentive for a book offering Thorne's own expert interpretation of what is on the screen, with the added benefit that his account of the background topics is an almost-understandable tour through science on the edge of reality.

The most intriguing aspects, for our purposes here, is Thorne's exposition of current spacetime theory explored by Lisa Randall (b. 1962, and a Harvard professor) and Raman Sundrum (b. 1964, and Distinguished University Professor at the University of Maryland). This proposes, in the version he adopts, not just the four dimensions of the universe we inhabit (considered as a membrane, or "brane") but also a larger realm within which it is embedded, the "bulk." This meta-universe has three space dimensions but only a single time dimension, like that of our brane, but also six additional spatial dimensions. If we could somehow climb free of the confinements of our brane world, we might be able to move in both space and time more freely than we can now.

This is enacted in the narrative of the movie, and for a full understanding of both the theory and its utilization in the plot one has to work through Thorne's ably illustrated book. But viewers of *Interstellar* will recall that on a climate-ruined Earth, a group of former NASA experts and astronauts construct a large starship capable of traveling to the vicinity of Saturn, where a bulk object has been found protruding into our conventional universe. The crew hopes to use it as a stargate, studying distant stars for habitable planets and seeking crucial quantum insights available only via a sojourn in the near vicinity of a distant black hole, Gargantua, before returning to Earth. The tremendous gravity of the black hole will, they know, dilate their time experience; this is the first kind of time travel. So hours there will be equivalent to several years on Earth—a heartbreaking sacrifice for astronaut Joe Cooper and his beloved and gifted 10 year old daughter Murph, who will be an old woman by the time he gets home.

For our purposes there is no need to go on at length about this tale that carries the science. The crucial discovery Coop makes is that entities exist in the bulk, where he finds himself trapped inside a four-dimensional tesseract (a higher-dimensional cube with eight faces, each a cube of the kind we are familiar with). Hard to visualize, but that was the Nolans' job—make Cooper's residence inside one of the faces of the tesseract comprehensible, and then solve the problem of how he can contact his daughter with no radio transmission possible between bulk and its enfolded brane.

This model of the greater cosmos argues that of the four known forces, only gravity can leak from one brane to another, via the bulk. The solution Cooper finds (spoiler!) is to manipulate the books in 10-year-old Murph's bedroom, using gravitational pulses to knock a sequence of books on the floor in a pattern that his brilliant kid will interpret as a message. Similarly, he learns to push on the "world-tube" of a watch, making its second-hand twitch in a Morse pattern that is stored in the bulk and played over and over until Murph understands and decodes, into a super-Einsteinian equation, the new quantum knowledge needed by humankind in order to escape their dying Earth.

Aside from this depiction of advanced physics, perhaps the most enchanting aspect is Thorne's explicit declaration that

> If wormholes are allowed by the laws of physics, then Einstein's relativistic laws permit transforming them into time machines. The nicest example of this was discovered [in 1988] by my close friend Igor Novikov, in Moscow, Russia. Igor's example… shows that a wormhole's transformation into a time machine might occur naturally, without the aid of intelligent beings. (266)

As noted previously, this might be undone in a literal flash when photons and other force particles cycle from one wormhole mouth to the other in a crescendo of amplifying energy. Thorne notes that he and his postdoc thought for a while that "the explosion was too weak to destroy the wormhole. Stephen Hawking convinced us otherwise" (268). However—possibly good news—

these educated guesses apply only to a universe with no bulk. If Einstein's model functions in the bulk, time machines might be saved even decades after he was first tempted by the equations to abandon them to his imagination:

> Lisa Randall, Raman Sundrum, and others have extended [Einstein's] laws into the fifth-dimensional bulk by one simple step: adding a new dimension to space. That extension proceeds mathematically in a straight-forward and beautiful manner, which makes us physicists think we may be on the right track. (269)

There are further complications, inevitably (perhaps, for example, even in the bulk nothing can go backward in time), but the game remains afoot.

A different way to approach the question of reversed time, a kind of elaboration of John Wheeler's and Dick Feynman's prankish guess that antimatter is just time-reversed matter of the usual kind, was developed by now-Emeritus Professor John G. Cramer (b. 1934) of the University of Washington in Seattle. His approach, dubbed the Transactional Interpretation (TI) of quantum physics and described in detail in *The Quantum Handshake* (2015), requires that quantum events comprise both a "retarded" wave function (moving into the future) and an "advanced" wave function (which moves in the other direction, reversed in time, carrying negative energy). "The Transactional Interpretation describes quantum wave functions traveling in both time directions in an even-handed and symmetric way" (161). This twofold model is not in itself new, but previously mathematicians and physicists have mostly ignored the counter-intuitive "advanced" wave function as a meaningless side effect of the calculations.

Cramer pointed out that they can be understood as an "offer" wave and a "confirmation" wave which interact to yield a probability that becomes realized when it is squared. Experiments to follow this interpretation into laboratory time machine territory have failed, so far. His time-symmetric conjectures are explored in detail in *The Quantum Handshake*, where he "pictures a transaction as emerging from an offer-confirmation handshake as a four-vector standing wave normal in three-dimensional space with endpoints at the emission and absorption verticies" (181). And, importantly, he notes: "these advanced time-running-backwards effects are limited to just the formation of time-forward transactions and are never allowed to produce 'advanced effects' that would violate cause-and-effect" (xii). That's depressing news for the Time Machine aficionado, but these are still early days.

Cramer's original TI has been critiqued and modified in several substantial studies by Dr. Ruth E. Kastner, a "possibilist" philosopher. Her approach is described by Cramer as "an alternative account of transaction formation in which the formation of a transaction is not a spatiotemporal process but one taking place on a level of possibility in a higher Hilbert space rather than in 3+1-dimensional spacetime" (181). Her most recent book is *Adventures in Quantumland: Exploring Our Unseen Reality* (2015), an attempt to combine relativistic and quantum methods within the TI framework.

In our present investigation of science-meets-time-travel, it should be mentioned that Cramer's TI has been given reader-friendly expositions in the sf magazine *Analog* in his regular "Alternate View" pop science series. Cramer commented in those pages:

> There are SF possibilities in the transactional interpretation. Advanced waves could perhaps, under the right circumstances, lead to "ansible-type" [faster than light] communication favored by [Ursula K.] LeGuin and [Orson Scott] Card and to backwards in time signaling of the sort used in [Gregory] Benford's *Timescape* and [James P.] Hogan's *Thrice in Time*. There is also the implication implicit in the transactional interpretation that Possibility does not become Reality along that sharp knife-edge that we call "the present". Rather, Reality crystallizes along a much fuzzier boundary which stitches into both future and past, advancing somehow in a way which defies sharp temporal definition.[5]

Cramer does not, however, project a future and past swarming with time machines riding the rails of advanced and retarded waves. In a similar vein, he dismissed one classic mid-twentieth century science fictional notion: "modern hard-SF has largely abandoned teleportation as a concept that has more to do with fantasy and parapsychology than with real science" (152). He does not mean the kind of real science currently fashionable among quantum physicists who transfer the state of one particle across space and onto another, but rather the handwavy kind familiar from Alfred Bester's *The Stars My Destination* (1956), Algis Budrys' *Rogue Moon* (1960) and, best known of all, *Star Trek's* "transporter."

(In my view it is a pity he dismisses parapsychology as equivalent to exhausted bad-science ideas, because the strong empirical evidence recorded by trained scientists such as those involved with the Star Gate program suggests that we do know that the future can modify the past, and that high-probability futures can be known non-inferentially in advance of their occurrence. Very possibly this body of work will provide a clue to how time machines might function, and within what constraints, once suitable formal theories arrive.)

---

[5] See, for example, https://www.npl.washington.edu/AV/altvw16.html

Another, on "quantum telephones, is https://www.npl.washington.edu/AV/altvw48.html

We find ourselves poised, then, with movie advisor and Nobelist Kip Thorne, between the frontiers of advanced science and the wilder imaginings of science fiction writers. Let us turn now to those fictional explorations of time travel, from H.G. Wells's *The Time Machine* six years prior to the start of the twentieth century in 1901, all the way through until today, where traveling in time has become a routine trope even in literary novels that dare not admit their links to "genre sci fi."

Part II

Time Machine Time

5

The First Half Century (and a Bit)

1895, *The Time Machine: An Invention*, H.G. Wells

In 1888, the brilliant working class English youth Herbert George Wells was 21 years old when he invented the notion of a powered machine that might convey its inventor into the future or the past and back home again. The serialized but unfinished short story where he displayed this idea was titled *The Chronic Argonauts* (published in *The Science Schools Journal* of the Royal College of Science) and later reviled by Wells for its amateurish crudity. He even "went to the length of buying up and destroying the back issues of the *Science Schools Journal* where its three instalments appeared…" In the next 7 years, as he developed as a writer of both fiction and what we now call "pop-science," Wells toyed with this fecund theme and finally published its first mature expression, the short novel (just 32,600 words in length, so strictly a novella) titled *The Time Machine*.

In the latter work, the inventor is identified only as "The Time Traveler" ("Traveller" in the British original), clearly a prosperous inventor who works in the well-stocked laboratory in his own upper middle class London home. The earlier tale, with its maladroit and now faintly comical title, is closer to Gothic horror. Its sinister genius, Dr. Moses Nebogipfel, deems himself "anachronic," born before his proper futuristic time. The Project Gutenberg release of this out-of-copyright story clarifies that unpleasant and almost unpronounceable name: "'Nebogipfel' is a linguistic hybrid consisting of the Russian word for 'sky' or 'heaven' (*nebo*) and the German word for 'peak' or 'summit' (*Gipfel*)." Dr. Nebogipfel looks faintly demonic:

© Springer Nature Switzerland AG 2019
D. Broderick, *The Time Machine Hypothesis*, Science and Fiction,
https://doi.org/10.1007/978-3-030-16178-1_5

He was a small-bodied, sallow faced little man, clad in a close-fitting garment of some stiff, dark material…. His aquiline nose, thin lips, high cheek-ridges, and pointed chin, were all small and mutually well proportioned; but the bones and muscles of his face were rendered excessively prominent and distinct by his extreme leanness. The same cause contributed to the sunken appearance of the large eager-looking grey eyes, that gazed forth from under his phenomenally wide and high forehead. It was this latter feature that most powerfully attracted the attention of an observer. It seemed to be great beyond all preconceived ratio to the rest of his countenance. Dimensions, corrugations, wrinkles, venation, were alike abnormally exaggerated. Below it his eyes glowed like lights in some cave at a cliff's foot. It so over-powered and suppressed the rest of his face as to give an unhuman appearance almost, to what would otherwise have been an unquestionably handsome profile. The lank black hair that hung unkempt before his eyes served to increase rather than conceal this effect, by adding to unnatural altitude a suggestion of hydrocephalic projection: and the idea of something ultra human was furthermore accentuated by the temporal arteries that pulsated visibly through his transparent yellow skin.

It can be no accident that this exaggerated skull, with its implication of either serious brain damage (hydrocephalus) or enhanced mental capacity, recurred in Wells's notable essay "The Man of the Year Million" (1893). It is a foreshadowing of the unsettling appearance of the superman in Olaf Stapledon's *Odd John* (1935) and other novels projecting a future of multiplied intelligence and emotional distance in a further evolved trans-humanity. Curiously, this is exactly what would *not* be proposed in the final shaping of this new device in *The Time Machine*, where the Traveler will encounter our debased descendants in the eight-hundredth century, the Eloi and Morlocks.

For the purpose of this book, though, we can ignore most of what makes these two works literary fiction: the Welsh landscape and its superstitious residents of Llyddwdd in *Chronic Argonauts*, the cod-Darwinian deep future of its amplified sequel. What makes these works relevant for our purpose is their account of time as a dimension no less traversable than the three spatial dimensions with which Wells's century was familiar. Here is Dr. Nebogipfel's account, which is already as well thought-out as that provided by the Traveler, published nearly two decades in advance of Albert Einstein's breakthrough mathematical insight into the nature of time as a fourth dimension, one leg of the fourfold substrate of physical reality now known as spacetime:

Has it never occurred to you that no form can exist in the material universe that has no extension in time?…Has it never glimmered upon your consciousness that nothing stood between men and a geometry of four dimensions—length, breadth, thickness, and duration—but the inertia of opinion...?

"When we take up this new light of a fourth dimension and reexamine our physical science in its illumination," continued Nebogipfel, after a pause, "we find ourselves no longer limited by hopeless restriction to a certain beat of time—to our own generation. Locomotion along lines of duration—chronic navigation comes within the range, first, of geometrical theory, and then of practical mechanics. There was a time when men could only move horizontally and in their appointed country. The clouds floated above them, unattainable things.... Speaking practically, men in those days were restricted to motion in two dimensions; and even there circumambient ocean and hypoborean fear bound him in.... Then man burst his bidimensional limits, and invaded the third dimension.... And now another step, and the hidden past and unknown future are before us. We stand upon a mountain summit with the plains of the ages spread below."

The notion of a fourth dimension was already well established in Wells's day, but as an extra *spatial* parameter, a possible direction through space at right angles to length, breadth and height or depth. Wells argued that duration, or extension in time, deserved to be accounted a dimension in its own right—and, once that were granted, perhaps one that could be negotiated or traversed just as the three spatial dimensions can permit motion forward or back, sideways, and up or down given suitable equipment (whether legs, submarines or aircraft). Why not, then, a machine allowing rapid transport to next year or perhaps a future age, or reversing into yesterday or even the depths of the past?

Throughout the subsequent century or so, the idea of time as a dimension became commonplace, although mostly as a device for fantasy or science fiction comics where the core meaning of "dimension" is confused with "world" or "variant universe." Famously, Rod Serling's television program invited viewers into *The Twilight Zone*,

a fifth dimension, beyond that which is known to man. It is a dimension as vast as space and as timeless as infinity. It is the middle ground between light and shadow, between science and superstition, and it lies between the pit of man's fears and the summit of his knowledge. This is the dimension of imagination.

Probably Serling meant "as timeless as eternity," but his metaphor for unbounded imaginative exploration, with all its risks and joys, is a *place*, a *location*. But really "dimension" is a kind of measurement, not a world other than the one we inhabit. In the *Superman* mythos, the Kryptonian superhero's whimsical and sometime malicious pest is Mister Mxyzptlk, who dwells in another version of the "fifth dimension" but visits ours by what amounts to magical means. Being a denizen of "another dimension," he is able to deform or ignore the restrictions of our customary three or maybe four dimensions. On the face of it, this is as silly as claiming that by climbing a ladder or a tall mountain we "enter the dimension of height," or by running along a street we are "transported into the dimension of length." Still, in a sense this confusing usage does reflect possibilities opened if it proves to be true that our reality is embedded in a fourth or higher *spatial* dimension—perhaps the "bulk" predicted by string theory and discussed in the previous chapter.

A famous analogy developed by the mathematicians Edwin A. Abbot and Charles Howard Hinton examined how a visitor from our more commodious spacetime would appear to those dwelling in a two dimensional cosmos. In Abbot's *Flatland* (1884) which might be imagined as an extended thin surface like a flattened bubble, a beach ball from our universe falling through the flat land would be seen first as a dot, then as an increasingly wider circle, until its circumference passed through, at which stage the circle would diminish to a point and vanish. Wells drew upon this conceit in his early *jeu* "The Plattner Story" (1887), in which luckless Gottfried Plattner slipped into a "strange Other-World" with four spatial dimensions and returned left-handed and with his internal organs reversed. If this actually happened, he'd have starved to death quite soon, because many of the foods we eat are built from either right-handed molecules or their reversed left-handed versions. Our digestive system is prepared for one kind but swiftly rejects the other. Wells was unaware of these chemical mirror-image mysteries of enantiomorphs (Greek for "opposite forms").

Today, advanced mathematical physics surmises the existence of more than just a fourth and perhaps fifth dimension needed to explain this quantum and relativistic cosmos. The consensus in string theory favors no fewer than ten dimensions, including time (seen as one-directional except at micro-levels) and the three spatial directions where increasing entropic disorder does not prevent motion in any of the usual directions. In Wells's pre-Einsteinian universe, retrocausation and hence time travel to the past is not forbidden, just difficult to master—just as flight required balloons or powered aircraft. Can the analogy of flight be pressed further by supposing that time's dimensionality does mandate a vaster reality, one where objects (and people) can be rotated out of the one-way "river of time" we daily experience and diverted into a "bulk" extended universe where the limitations of our world are surpassed?

Doing something along those lines does seem necessary if the Time Machine Hypothesis (escape, that is, from the inviolable dominion of time) is to be taken seriously. It's all very well for the time machine builder to rejoice in standing upon a mountain summit with the plains of the ages spread below, but how does he get there? What energy source powers his machine, and what is it made of? What effects would it have upon its local environment? Why does it vanish from sight and touch as it flits off into the future or the past?

Consider that last question seriously. Unless the Time Traveler's ride does rotate itself into a realm which can't be detected by senses evolved for only three dimensions of space, as a balloon floating through Flatland simply disappears without leaving a trace, its accelerated journey should create dangers for anyone in its vicinity. Wells was aware of this implication, and deftly avoided it with the skill of a practiced card sharp. In *The Time Machine*, its inventor displays a diminutive but working mock-up of the craft, and activates it in front of his dinner guests. The Psychologist, seeing it disappear, says, "I presume that it has not moved in space, and if it travelled into the future it would still be here all this time, since it must have travelled through this time." Wrong, he is told.

> *You* can explain that. It's presentation below the threshold, you know, diluted presentation.
> "Of course," said the Psychologist, and reassured us. "That's a simple point of psychology. I should have thought of it. It's plain enough, and helps the paradox delightfully. We cannot see it, nor can we appreciate this machine, any more than we can the spoke of a wheel spinning, or a bullet flying through the air. If it is travelling through time fifty times or a hundred times faster than we are, if it gets through a minute while we get through a second, the impression it creates will of course be only one-fiftieth or one-hundredth of what it would make if it were not travelling in time. That's plain enough." He passed his hand through the space in which the machine had been. "You see?" he said, laughing.

Really? It is like looking at a rapidly spinning bicycle wheel, or a fan blade, where the passage of light is blocked for only a fraction of a second, so (he is claiming) you can easily reach into the now-invisible wheel or blade. He fails to add: Until it chops your fingers off.

What's more, some of the most impressive and engaging images of travel in such a device would seem to be ruled out by this *faux* explanation. Gazing out into the world from his accelerating machine, this is what Time Traveler sees:

> The laboratory got hazy and went dark. Mrs. Watchett came in and walked, apparently without seeing me, towards the garden door. I suppose it took her a minute or so to traverse the place, but to me she seemed to shoot across the room like a rocket. I pressed the lever over to its extreme position. The night

came like the turning out of a lamp, and in another moment came tomorrow. The laboratory grew faint and hazy, then fainter and ever fainter. Tomorrow night came black, then day again, night again, day again, faster and faster still. An eddying murmur filled my ears, and a strange, dumb confusedness descended on my mind.

I am afraid I cannot convey the peculiar sensations of time travelling. They are excessively unpleasant. There is a feeling exactly like that one has upon a switchback—of a helpless headlong motion! I felt the same horrible anticipation, too, of an imminent smash. As I put on pace, night followed day like the flapping of a black wing. The dim suggestion of the laboratory seemed presently to fall away from me, and I saw the sun hopping swiftly across the sky, leaping it every minute, and every minute marking a day. I supposed the laboratory had been destroyed and I had come into the open air. I had a dim impression of scaffolding, but I was already going too fast to be conscious of any moving things. The slowest snail that ever crawled dashed by too fast for me. The twinkling succession of darkness and light was excessively painful to the eye. Then, in the intermittent darknesses, I saw the moon spinning swiftly through her quarters from new to full, and had a faint glimpse of the circling stars. Presently, as I went on, still gaining velocity, the palpitation of night and day merged into one continuous greyness; the sky took on a wonderful deepness of blue, a splendid luminous color like that of early twilight; the jerking sun became a streak of fire, a brilliant arch, in space; the moon a fainter fluctuating band; and I could see nothing of the stars, save now and then a brighter circle flickering in the blue.

The landscape was misty and vague. I was still on the hillside upon which this house now stands, and the shoulder rose above me grey and dim. I saw trees growing and changing like puffs of vapour, now brown, now green; they grew, spread, shivered, and passed away. I saw huge buildings rise up faint and fair, and pass like dreams. The whole surface of the earth seemed changed—melting and flowing under my eyes.

Delightful or even alarming as all this is, then since enough light filters into his flickering presence (even though "faint and hazy, then fainter and ever fainter"), by the same token his machine must appear briefly in a kind of drawn-out quantum jumping fashion. Too bad for anyone who chances to walk into it at such instants, or to building construction that happens to be erected on its former site.

Could it be (although Wells did not have the information available to conjecture this) that his machine somehow warps local spacetime into a sort of closed knot, akin to a black hole? We know that the immense gravity of a star-sized black hole slows time to a crawl in its vicinity (one of the driving plot devices in the movie *Interstellar*). But the mass of a human or two inside a light-weight Time Machine would not generate a strong enough gravity field inside a closed timelike curve to lift a mosquito off the dashboard, let alone to arrest aging and decay over millions of years.

What does this remarkable device look like? The description remains somewhat consistent from *The Chronic Argonauts* on. Here is how Dr. Nebogipfel and his companion and their craft appear as they return from an excursion into the future:

> It was solid, it cast a shadow, and it upbore two men. There was white metal in it that blazed in the noontide sun like incandescent magnesium, ebony bars that drank in the light, and white parts that gleamed like polished ivory. Yet withal it seemed unreal. The thing was not square as a machine ought to be, but all awry: it was twisted and seemed falling over, hanging in two directions, as those queer crystals called triclinic hang; it seemed like a machine that had been crushed or warped; it was suggestive and not confirmatory, like the machine of a disordered dream.

Here is the toy machine built in his lab by the Time Traveler:

> The thing [he] held in his hand was a glittering metallic framework, scarcely larger than a small clock, and very delicately made. There was ivory in it, and some transparent crystalline substance.... You will notice that it looks singularly askew, and that there is an odd twinkling appearance about this bar, as though it was in some way unreal.... Now I want you clearly to understand that this lever, being pressed over, sends the machine gliding into the future, and this other reverses the motion. This saddle represents the seat of a time traveler.

The full-sized machine is structurally similar, and light enough that when it tumbles over it can with some trouble be hoisted back up by one person and dragged some distance without damage. But no account is given of its mode of operation or power supply. In an authorized sequel by Stephen Baxter, *The Time Ships*, published for the novel's centenary in 1995, the energy source is given as "Plattnerite." This glowing green and possibly radioactive crystal is the kind of imaginary material or fuel sometimes referred to as "unobtainium." Perhaps our capacity to build and travel in a time machine must await our discovery of such a substance.

1919, "The Runaway Skyscraper," Murray Leinster

Born in Virginia a year after the publication of H.G. Wells's *The Time Machine*, William Jenkins (1896–1975) wrote early and late in a variety of voices and genres, often under the Leinster pseudonym (pronounced "Lenster"). He was an ingenious if stylistically unadventurous story-teller—he seems to have invented the notion of alternative histories in "Sidewise in Time" (1934), and predicted the internet in "A Logic Named Joe" (1946). So perhaps it was inevitable that he should have seen possibilities in time travel beyond the very limited one-person explorer in *The Time Machine*. We know this, because the "young and budding" engineer protagonist Arthur Chamberlain asks his demure, terrified stenographer Estelle Woodward, during what we might call a catastrophic real estate accident,

> [H]ave you ever read anything by Wells? *The Time Machine*, for instance?
> [S]he shook her head.
> I don't know how I'm going to say it so you'll understand, but time is just as much a dimension as length and breadth. From what I can judge, I'd say there has been an earthquake, and the ground has settled a little with our building on it, only instead of settling down toward the center of the earth, or side-wise, it's settled in this fourth dimension.
> "But what does that mean?" asked Estelle uncomprehendingly.
> If the earth had settled down, we'd have been lower. If it had settled to one side, we'd have been moved one way or another, but as it's settled back in the Fourth Dimension, we're going back in time.
> Then—
> We're in a runaway skyscraper, bound for some time back before the discovery of America!

Even as he offers his rather condescending explanation, their New York building, the Metropolitan Tower, is indeed plunging backward through time, displaying through its office windows some of the same accelerated effects the Time Traveler had observed, but not quite as dizzyingly. What they see is history in reverse:

> There was hardly any distinguishing between the times the sun was up and the times it was below now, as the darkness and light followed each other so swiftly the effect was the same as one of the old flickering motion-pictures.
> As Arthur watched, this effect became more pronounced. The tall Fifth Avenue Building across the way began to disintegrate. In a moment, it seemed, there was only a skeleton there. Then that vanished, story by story. A great cavity

in the earth appeared, and then another building became visible, a smaller, brown-stone, unimpressive structure.

With bulging eyes Arthur stared across the city. Except for the flickering, he could see almost clearly now.

He no longer saw the sun rise and set. There was merely a streak of unpleasantly brilliant light across the sky. Bit by bit, building by building, the city began to disintegrate and become replaced by smaller, dingier buildings. In a little while those began to disappear and leave gaps where they vanished.

Arthur strained his eyes and looked far down-town. He saw a forest of masts and spars along the waterfront for a moment and when he turned his eyes again to the scenery near him it was almost barren of houses, and what few showed were mean, small residences, apparently set in the midst of farms and plantations.

His watch has exploded which Arthur understands is due to the mechanism running backward until the spring is intolerably tightened, and breaks. He fails to explain why this counter-clock retroactivity has not reversed their own metabolic and cortical processes, or why the great coal-fueled generator in the bowels of the falling skyscraper is still providing ample electricity for lighting and elevators rather than grinding to a halt or transforming the heat in the building back into pre-burned coal.

Leinster's narrative arc follows a blend of themes that decades later would find fictive expression in *Lord of the Flies* (marooned, polite schoolboys tend to regress to brutish survivalism), *The Admirable Crichton* (working class butler, disdained and ignored by his wealthy socialites, whips them all into shape on a desert island after their luxury yacht sinks) and *MacGyver* (faced with serious dangers, a handsome young man improvises clever repairs and gains freedom for himself and his companions). Arthur's analysis of the problem is withheld from the reader for a time, because this tale is an adventure story for *Argosy and Railroad Man's Magazine*, not a metaphysical study of the fourth dimension as a rail-less road to other times.

With the aid of the gray-haired president of the skyscraper's ground floor bank, Arthur harries the two thousand lost hapless souls of the building into a measure of solidarity when the structure finally settles thousands of years earlier in an all but virgin forest near the sea, the cleared half-acre home of a small village of native people who have not yet mastered iron-working. This concord does not last long, as hungry men express their anger and frustration by smashing into a canteen and rifling the food. Soon enough, though, Arthur and the bank manager organize a chastened rabble into setting aside a space for the women to sleep unmolested on scavenged cushions. Teams of anglers seek fish while hunters bear their few arms into the woods in search of game.

The tense communal *glasnost* is saved when vast numbers of passenger pigeons hurl themselves into the external walls and lighted windows, dropping stunned and ready for dinner. And so on.

The explanation is somewhat obscure to the non-engineer of 100 years later, but it amounts to this: an earthquake has unsettled the vast mass of the skyscraper, still positioned on its huge concrete piles that reach deep into bedrock; in the center of one of those piles is found a hollow tube "originally intended to serve as an artesian well." An underlying geyser or hot spring is now blocked. Arthur and his sweating team drill holes, "soap" the blockage (don't ask), and finally the flow is returned to a suitable status and the skyscraper settles. Why? In this case, alas, there is no point in asking, because, Leinster informs us candidly, "For the fact that this 'sinking' was in the fourth direction—the Fourth Dimension—Arthur had no explanation."

It was still too soon, in 1919, to conjure explanations from advanced physics, even though Einstein had by then published not only on Special Relativity but also on General Relativity. We can't really blame Leinster, any more than we rebuke Wells whose Time Traveler never makes any attempt to present a wiring diagram for his device of nickel, ivory and rock crystal, let alone a theorized basis for its ability to move outside conventional duration.

What we can guess in advance is that the Tower will return to its time and place of origin—to the exact second of its departure, as it turns out—and that Arthur and Estelle will kiss while "the nights and days, the winters and summers, and the storms and calms of a thousand years swept past them into the irrevocable past," as "three generations were born, grew and begot children, and died again!"

1938/rev 1952 *The Legion of Time* Jack Williamson

Jack Williamson (1908–2006) was one of the founders of modern science fiction, and lived to see it change from coarse adventure tales to genuinely sophisticated literature. He was himself in the early days as pulpy a writer as the others, but by the mid century he had gained BA and MA degrees in English, and later a PhD on H.G. Wells, teaching literature for many years. He wrote influential tales about terraforming and antimatter, as well as horror fiction. His time travel/alternate reality novel *The Legion of Time* was perhaps the first to combine headlong adventure with explicit borrowings from quantum mechanics long before popular science magazines bought non-specialists the highly counter-intuitive news of quantum superposition.

Denny Lanning, an 18 year old Harvard student who later becomes a war journalist, is contacted mentally by two opposed beautiful women from the future, or rather futures, and on this basis must choose whose possible universe is chosen for realization. Lethonee, a gorgeous pale and silver-voiced advocate for Jonbar, a future utopian city, warns him at moments of impending death, but at the cost of other lives. Golden-voiced Sorainya, clad in crimson armor, represents the bleak dystopian Gyronch, a world of malign bio-engineered ants, and explains that he must die before it is possible to transport him to either of the competing realities. The way to implement one's choice of tomorrows is literalized by Lethonee:

> The world is a long corridor, from the beginning of existence to the end. Events are groups in a sculptured frieze that runs endlessly along the walls. And time is a lantern carried steadily through the hall, to illuminate the groups one by one. It is the light of awareness, the subjective reality of consciousness.
>
> Again and again the corridor branches, for it is the museum of all that is possible. The bearer of the lantern may take one turning, or another. And always, many halls that might have been illuminated with reality are left forever in the dark (19–20)

Denny, she explains, is destined, for a little time, to carry the lantern.

(Curiously, and surely not by accident, Poul Anderson's 1965 time travel novel of competing futures is titled *The Corridors of Time*. There, time machines are very long subsurface tunnels that link decades and centuries by a species of hovering sled—giving the term "passage of time" a new meaning.)[1]

Of course Denny lusts after the Satanic lure of the Queen of Gyronch even as his better angels urge him to embrace Lethonee, and her vision. He more or less comes to his senses when a chronoscope invented by his old college pal and now professor Wil McLan displays the wretched ruin of the Earth wrought by Sorainya, her slaves, and bestial squads of anthropoid ants. Lanning is also dismayed by the ruined state of McLan's body after being subjected by her to appalling torture for 10 years. Due to a sort of geodesic temporal relativistic effect, the professor is now some 40 years the senior of 28 year old Lanning, but not so enfeebled that he cannot pause for pages at a time to explain the

---

[1] What *is* a chance (and rather spooky) coincidence is the name of a pivotal young woman representing one of two possible futures in my 1982 novel *The Judas Mandala*: she is Sriyanie, rather too close to Sorainya for my comfort. However, her name was slightly modified from the given name of an Asian co-worker of my then-partner some forty years ago, when I had not yet read Williamson's novel.

physics of the situation. For today's readers who cannot abide bald data dumps of the Tell Don't Show kind, this and the abundant use of sobbing, exclamation points!, and so on make the book barely tolerable. For our purposes, though, it is handy (if handwaving) to have Williamson's model of time travel and contesting worlds presented.

McLan is especially remorseful because it was his love-sick obsession with the enticing Sorainya that drove him to spend years tracking her movements with his chronoscope, unaware until too late that this transaction of immense atomic forces strengthened her world at an ontological level while vitiating Jonbar's. There proves to be only one way to resolve this Gnostic contest between Good and Evil and that lies in a single moment of history: 5:49 pm, August 12, 1921, when an uneducated lad named John Barr, driving his two spotted cows along a dusty trail, bends to pick up one of two adjacent objects.

One is a broken piece of magnetized iron with a rusty nail clinging to it, which will set his curious mind on the path to science and eventually discovery of dynatomic tensors, or dynat. This is "a totally new law of nature, linking life and mind to atomic probability"—opening up "a tremendous new technology for the direct release of atomic energy, under the full control of the human will" (81). With this breakthrough, humankind creates the democratic beauty and achievements of the place named for him, Jonbar.

The other, a sparkly pebble, has no qualities to stimulate Barr's youthful mind deeper than to "toss the pebble in his hand, and throw it in his sling to kill a bird." Part of the secret of *dynat* is discovered instead, years after his death, by exiled Soviet engineer Ivor Gyros and a renegade Buddhist priest, who name it *gyrane* and use it to establish a fanatical and despotic empire that leads at last to Sorainya's "dark dynasty" and its entropic collapse (82). Lanning and the dying McLan, after a heroic fight in and out of the probabilistic hyperspace, best the foe at the cost of most of their "legion"—but happily the *dynat* phenomenon allows the dead to rise. In one final twist, the geodesic leading to Jonbar recovers as well, taking them home, and Lanning (rather luckily for all concerned) finds his beloved and angelic Lethonee now doppelgängered with her superposed foe:

> the golden voice of the [now slain] warrior queen had mocked him in her cry. And the ghost of Sorainya's glance glinted green in her shining eyes… neither Jonbar nor Gyronchi had ever actually existed. Divergent roads of probability stemmed from the same beginning, they were now fused into the same reality. "Yes, my darling." He drew them both against his racing heart… (102)

1941 "By His Bootstraps" Robert Heinlein

As we have seen, one of the persistent objections raised by philosophers and scientists to the possibility of time travel to the past is toxic time loops. These might retroactively interfere with events or conditions crucial to the traveler's history, perhaps even short-circuiting his or her own existence. Such loops, of course, entail logical *paradoxes*, always a strong hint that the events described are innately un-physical—which is to say, impossibly self-defeating. Nothing of this kind was hinted in H.G. Wells's own time machine tales, but it was quickly noticed as a piquant and entertaining trope for science fiction pioneers to play with, ringing all the feasible changes and tying all the likely causal knots. When the ingenious Robert Heinlein, invalided out of the Navy prior to the second world war, turned his hand to sf as a form of income, one of his early efforts was just such a story of temporal looping.

Published in John W. Campbell's *Astounding*, under the *nom de plume* Anson MacDonald, "By His Bootstraps" (1941) was quickly hailed as the classic exemplar of such unsettling time loops. It would retain that distinction until 1959, when Heinlein outdid his own cleverness with "All You Zombies" (*Fantasy & Science* Fiction), in which an orphan girl grows up to become her own mother *and* father, under the instigation and tutelage of a much older version of himself.

But *is* the twisted causal sequence in "By His Bootstraps" a paradox, or just a puzzle with some of the working parts hidden from the reader? Might it show us some of the ways a self-consistent series of events could, in principle, save time travel from the fate of logical extinction?

Here is the narrated sequence of events:

Twenty-two year old Bob Wilson, a doctoral candidate and pre-war slacker, has left completion of his thesis to the very last minute. That is, our focus is drawn immediately to time as a factor in his life. Fueled by coffee and desperation, he sits at his desk for thirteen hours, smoking fifty-two cigarettes, and writing 7000 words toward his "Investigation into Certain Mathematical Aspects of a Rigor of Metaphysics." By what might seem a strange coincidence, or worse still a lazy prolepsis by the author, he types:

> A case in point is the concept "Time travel." Time travel may be imagined and its necessities may be formulated under any and all theories of time, formulae which resolve the paradoxes of each theory. Nevertheless, we know certain things about the empirical nature of time which preclude the possibility of the conceivable proposition. Duration is an attribute of consciousness and not of the plenum. It has no *Ding an Sich* [thing in itself] Therefore—
>
> The key of the typewriter stuck…

While Bob has been typing this moderately sophisticated academic prose, a wormhole mouth (as physicists would come to call such speculative phenomena decades later) has opened behind him, from which steps a slightly later version of himself, who says dismissively "Don't bother with it… It's a load of utter hogwash, anyway." Shortly an even later version of himself arrives, they all fight, and the initial Bob is knocked through the Time Gate some 30,000 years into the future.

These tangles and repetitions are fairly predictable to today's sf readers, nearly 80 years after Heinlein wrote it, and somewhat less convincing than it must have seemed at the time. (For a start, we have now had many decades of stories, books and movies about the topic to prime us for the troubles Bob faces in pristine ignorance, despite his apparent familiarity with early academic thinking on time travel.) Bob finds himself in the Hall of the Gate in the High Palace of Norkaal, ordered about by a tough-minded fellow he fails to recognize as his 10-year older self, prematurely white-haired and bearded, identifying himself as Diktor (Dictator? Director? Doctor? We never find out). Somewhat like Wells's Time Traveler, Diktor is surrounded by the equivalent of beautiful, passive Eloi with no Morlocks to pester or eat them, although he treats them as slaves and offers a lovely woman to Bob as a sexual gift. He sends Bob (now, we realize, Bob_2, back to fetch Bob_1, although one would expect Bob_n as Diktor to recall the difficulties this had engendered in his own past. (These subscripts are not present in Heinlein's text, of course.)

After more convolutions, Bob_3 (as he has become by now) is introduced to the steering device of the time machine allowing all this temporal gallivanting, four control spheres arranged as the vertices of a tetrahedron. The base trio command motion in three spatial dimensions of what amounts to a time viewer, while the top-most allows motion backward or forward in time. Stealing a handwritten notebook of vocabulary and escaping from Diktor's scrutiny (although, we realize, Diktor remembers doing exactly this 10 years earlier by his accounting of time, and permits the necessary escape), Bob explores the huge structure built long ago by vanished non-humans who called themselves "the Forsaken." He suffers extreme and long-lasting trauma in an encounter with no more than the time viewer image of one such. "He had been flicked with emotions many times too strong for his spiritual fiber and which he was no more fitted to experience than an oyster is to play a violin."

Recovering, and keeping an eye peeled for Diktor, Bob settles among the quasi-Eloi, learns from the notebook that "Diktor" is a title rather than a name and appropriates it for his own use, directs some improvement in the lives of his passive subjects, and finally works out who he is. He copies the now tattered and worn notebook into a clean version and destroys the original, understanding only belatedly that the new notebook must be the same one he copied—in which case, who did the research annotated helpfully therein? It's no part of his own memory.

Here we find a genuine paradoxical loop, rather gratuitously introduced by Heinlein for its bafflement value. It was strictly unnecessary to the novella, since Bob/Diktor could have learned the far future language by oral interactions with the docile Eloi. But it gives the author the opportunity to provide this frisson: "The physical process he had all straightened out in his mind, but the intellectual it represented was completely circular." It is the same paradox as a student going forward in time and copying the published text of his future discoveries and then releasing them without having to do the research work that underpinned them.

This miraculous appearance of new information can only be justified if there is a multiplicity of parallel worlds, in some of which the research has indeed been done and published, and the method used to filch this knowledge is actually a cross-time device, not strictly a time machine (a proposition argued by quantum physicist David Deutsch). This does not occur to Bob, nor perhaps to Heinlein, although the notion of historically variant worlds had already been introduced in sf by such innovators as Jenkins/Leinster in "Sidewise in Time"(1934) and he used a version of it himself in "Elsewhen" (also in *Astounding*, 1941).

It is arguable that part of the mystery is enfolded in the now-vanished Forsaken, who presumably built and then abandoned the mighty High Palace of Norkaal; they are the likely source of the Gate and its time viewer apparatus. Or perhaps the true source was another even more mysterious species, whether organic, robotic or "spiritual," that preceded those and left them "forsaken." In either case, it might be that an automated system in the far future activated a search through time and space in search of a young human thinking deeply about time travel, and opened a Gate portal that allowed a rosette of retrocausal actions leading to Diktor and thus back to Bob_1 and perhaps the recovery, beyond the end of Heinlein's story, of the quasi-Eloi. What would be their motive? We cannot know, or perhaps even guess. Such are the puzzles of the deep future and its denizens, which leave us with "as much chance of understanding such problems," as Bob reflects at last, "as a collie has of understanding how dog food gets into can."

1942 "Recruiting Station" aka *Masters of Time,* aka *Earth's Last Fortress* A.E. van Vogt

In 1939, Canadian Alfred Elton van Vogt (1912–2000, and generally known as Van) burst into the genre sf scene at a pivotal point in its transition from largely low-grade adventure fodder to what became "Modern science fiction," under the guidance of John W. Campbell. This convulsive transition is often held to mark the end of slovenly writing ("his brain staggered, literally") and mostly unschooled mock-science and a new emphasis on more mature, informed and competent story-telling.

There's some truth in that evaluation, but it didn't prevent Campbell from publishing a significant amount of abject drivel, especially from the like of L. Ron Hubbard and van Vogt. What justified the feverish and often inane stories by van Vogt, though, was the imaginative impact they produced in readers young and old, now diagnosed by scholars as due to its oneiric or dreamlike character. This was wedded to a genuine "sense of wonder" at the majesty of the cosmos and an expectation of future transformation, and often transcendence, in its human and posthuman inhabitants. As the Panshins observed in *The World Beyond the Hill* (1989), "'Recruiting Station' would be notable for its presentation of a future containing not just change upon change, but level upon level of possible human becoming" (502).

Despite Van's dreadful writing and frequent silliness, some mysterious essence seemed to rise from the pulpy paper of *Astounding* and other magazines, releasing a mood of entranced anxieties and exultations in its readers. It didn't matter that what you were reading made little or no sense, or that the emotions attributed to characters jagged about wildly in a single, heavily impastoed paragraph: this was the real deal, somehow. Attempting to deploy carefully rational explanations for the event-stream would have dragged the whole process to a halt, tipping the reader out of dream and back into dismal 1940s wartime tedium and terror. It was inevitable that time travel, and more generally a kind of large-scale temporal engineering, would seize Van's imagination. This narrative engine reached an early pitch in "Recruiting Station," a 31,000 word novella set in 1941–1944 and thereafter millennia into the future and, ultimately, by one trapped character's numb estimate, a quadrillion years. Meanwhile, we learn that colossal "time-energy" destroys and remakes the universe millions of times a second, multiplying realities in a profusion (ten billion recreations of the solar system every second) that not only anticipated the Many Worlds or Relative State hypothesis of Hugh Everett III but perhaps outstripped its extravagances.

Norma Matheson is a young woman at the end of her tether in 1941, nerving herself for suicide in a "broad, black river [that] gurgled evilly at her feet." Eleven years earlier, driven by ambition for a career of her own, she had declined an offer of marriage from the brilliant young scientist Jack Garson. Now her hopes of independence, "ambitious for advancement in the important field of social services," have faded; she is a victim, as it were, of time and its freight of entropy. Sitting beside the water, she is addressed by Dr. Lell, a tall, dark-skinned man of uncertain ethnicity with epicanthic folds above his eyes and a faint accent. Lell knows her name and can plainly read her mind; an immortal, he proves to be from the future, one of the pompously self-named Glorious, "masters of time [who] live at the farthest frontier of time itself, and all the ages belong to us." He offers Norma employment and free accommodation in his bogus operation supposedly in support of "the Calonian cause." (Van's lightly-disguised version of Catalonia; in our real history, American volunteers fought in the Spanish civil war, but their leftist Abe Lincoln Battalion would be illegal, and later condemned by US authorities as "premature anti-fascists").

In her misery and helplessness, Norma accept his offer, perhaps augmented by advanced hypnosis, and recruits these luckless men to a condition of servitude in a vast futuristic war against the Planetarians. (Lell rants against his foes in pulp diction: "We shall win this war, in spite of being on the verge of defeat, for we are building the greatest time-energy barrier that has ever existed. With it, we shall destroy—or no one will win! We'll teach those moralistic scum of the planets to prate about man's rights and the freedom of the spirit. Blast them all!")

Despite the primitive writing, van Vogt's phantasmagoria offers an immense vista of human and posthuman command of time. The temporal instruments of the Glorious are not the one- or two-person homemade gadget of Wells's Time Traveler, but great engines of power. Even Norma's modest recruitment office has one in a spare room: "The machine crouched there, hugging the floor with its solidness, its clinging metal strength; and it was utterly dead, utterly motionless." It doesn't stay inert for long:

Incredibly, the machine was coming alive, a monstrous, gorgeous, swift aliveness. It glowed with a soft, swelling white light; and then burst into enormous flame. A breaker of writhing tongues of fire, blue and red and green and yellow, stormed over that first glow, blotting it from view instantaneously. The fire sprayed and flashed like an intricately designed fountain, with a wild and violent beauty, a glittering blaze of unearthly glory.

And then—just like that—the flame faded. Briefly, grimly stubborn in its fight for life, the swarming, sparkling energy clung to the metal.

It was gone.

Quite why a time engine requires such pyrotechnics is never explained, and it seems rather a waste of energy to no good purpose, but like the ravening special effects of "sci-fi" and fantasy movies more than half a century into Van's own future it is designed to thrill innocent readers, and succeeded in doing so. Norma, though, is horrified by this display of force and warns Lell that she means to alert the police. But she is now a "slave to the machine," and as she approaches an imposing police building she stumbles, weakening, and psychically ages into "a tall, thin, old, old woman." Lell takes her in charge and explains that his device can suck out her vitality or replenish it if she does what she is told. Here is a different variety of time control, far closer to magic than science although Norma eventually learns to tweak controls secreted in the key to the office and return her body to the fresh youthfulness of a 20 year old, as she'd been when she left Jack.

In the future, a murderous war of attrition is being fought above the Glorious city of Delpa between their destroyers (crewed by "depersonalized" troops from the past) and aerial battle cruisers of Planetarians. Beyond both, the inevitable third stealthed players intercede to aid both Norma and Jack in their separate captivity. These beings from the four hundred and ninetieth century know that if the Glorious find themselves at the edge of defeat, they can and will destroy the entire universe. Only the most indirect steps can be used to neutralize them:

> The reason we cannot use so much as a single time machine from our age is that our whole period will be in a state of abnormal unbalance for hundreds of thousands of years; even the tiniest misuse of energy could cause unforeseeable changes in the fabric of time energy, which is so utterly indifferent to the fate of men. Our method could only be the indirect and partially successful one of isolating the explosion on one of eighteen solar systems, and drawing all the others together to withstand the shock. This was not so difficult as it sounds, for time yields easily to simple pressures.

In this case, the simple pressures are employed by Norma and Jack, whose history is refigured to an apparent compliance with Lell's demands, thus creating the illusion for the Glorious that she spends the coming years as a true slave to the time machine. Finally, married to Professor Jack, she will have learned to master her implicit powers and "destroy the great energy barrier of

Delpa and help the Planetarians to their rightful victory." Easy when you know how, with good schooling.

Finally, then, as Norma sits beside the water where her earlier instantiation contemplated suicide, she watches Dr. Lell approach in the dark of the night, and van Vogt closes this tale of wild time contortions with one of the great and memorable phrases of science fiction: "Poor, unsuspecting superman!"

1946 "Vintage Season" by C. L. Moore

At an extreme remove in sensibility and literary tact from van Vogt's triumphalist word salads, Catherine Moore's classic story considers the possible use of time travel without providing the slightest idea of what its machinery looks like or how it works. Here is one of the first treatments of a now well-honed sf concept, time travel as a vehicle for ambitious tourists. It has often been anthologized, and appeared in one of the *Science Fiction Hall of Fame* volumes, although, as John Clute notes in *The Science Fiction Encyclopedia*, it shows an "earnest emotionality that was underappreciated" in the 1940s shortly after the end of the second world war.

Moore (1911–1987) here provides three startlements that, more than is usual in a book of this sort, might justify a stern SPOILERS! warning. Before we turn to them, it is worth noting that this novella first appeared in *Astounding* as by "Lawrence O'Donnell," and these days under the names of both Moore and her writer husband Henry Kuttner (1915–1958). Married in 1940, the Kuttners were enormously productive and diverse in their plots and approaches, not only inspiring each other but frequently using the "hot typewriter" collaborative method in which one writer begins a story then hands it over, sometimes in the middle of an unfinished sentence, for the partner to continue, and so on. It is now known, however, that Moore alone wrote this luminous story.

The first dazzlement is just the discovery by Oliver Wilson, young owner of a decrepit mansion, of the nature of the gorgeously dressed and oddly spoken trio who've rented the entire building for the month of May. Using the surname Sancisco, they are Omerie, a powerfully confident man, and two strikingly beautiful women, Kleph and Klia. They prove not to be visitors from an obscure nation (perhaps, a non-sf reader might speculate, in Latin America or the then-Communist realm behind the Iron Curtain, or even Mars) but are from the future. What, then, attracts them here to the late 1940s, and what possible interest in this particular old house has driven another woman of their enigmatic kind to offer three times the entire regular purchase price of the place? Oliver's somewhat shrewish fiancé Sue urges him to break the existing month's lease by driving the three early birds away; she wants to grab this gift horse before it trots away. Oliver is uncertain; he already has fallen more than half in love with lovely Kleph.

The second shock is his experiencing a devastating meteor impact that levels their unnamed city while a large group of the futuristic tourists watch with drugged excitement through upstairs windows as the broken ground roars, fires break out, buildings as far as the eye can see lie crushed and broken as tinder, and voices of lamentation can be heard bursting from the throats of the survivors. Probably Sue is dead in the ruins. This is the gruesome entertainment that has drawn these artistic connoisseurs.

The final dramatic coup, revealed in the final two words, is the Blue Death. This dreadful plague instantly infects Oliver and presumably most of the survivors and, in due course, maybe most of the world. Is this a disease carried from space with the "falling star"? Or, more horrifying, a plague spread unintentionally by the time tourists themselves (who bear healing inoculation scars), like the bacteria and viruses Europeans brought to the Americas and Australia that infected and slaughtered millions? Probably not the latter, we realize on reflection, since these travelers have recently visited Capri at the summer height of the reign of Emperor Augustus, and Chaucer's autumnal if plague-ridden Canterbury at the end of the fourteenth century (and plan to see Rome next, for the coronation of Charlemagne). Since these localities were not subjected to the Blue Death, the time travelers cannot be Typhoid Marys, merely protected against the Death—but their callous holiday in twentieth century misery is dying Oliver's last experience of this perfect month.

There is considerable theatric force in these calamities, but rather too much contrivance. At first we are told that these wanderers in time have arrived to enjoy a week of May historically unparalleled for its glorious weather, the "warm, pale-gold sunshine and the scented air" not for a moment interrupted by rain or cold. It seems they have been pursuing this radiant climatic moment, but of course it is the promised shattering shock of meteoric impact that whets their appetite. This ironic coincidence of benign sun and air with monstrous death falling upon the innocent is perhaps notable enough to explain why this place and time, out of all available chronology, attracts tourists—but the arrival of a lethal plague pandemic might seem, literally, overkill. It is a pleasing concatenation for these jaded visitors, but more than that. Their own century's greatest artist, Cenbe, who creates three dimensional symphonia of sound and light and darkness, music and image, is there with them, completing his great saga of emotional crisis.

History itself, of course, was the artist—opening with the meteor that forecast the great plagues of the fourteenth century and closing with the climax Cenbe had caught on the threshold of modern times. (679)

Moore's novella is not interested in the mechanisms of time travel but in the motivations of those who use it. Wells's Time Traveler was the scientist who developed a theory of time transport as well as the engineer who built the first such machine, but after his initial exploratory impulse he was driven not only by fear of his own death but solicitude for the child-like Weena and her Eloi companions menaced by cannibalistic Morlocks. With Weena burned to death in a fire of his own making, he rushed into the far future, to the imminent death of the Sun and the frozen Earth, in anguish, and planned a return (it seems) to repair the damage he had wrought. But we can never know if he succeeded in this quest.

Catherine Moore's tale drives this crux into a poignant and horrifying Grand Guignol, where disinterested spectators view cosmic horror and coolly make great art of it, in Cenbe's harrowing catalogue of pain. This was a crux arguably possible only in fiction of time voyaging, and it is remarkable that such a measure of emotional and artistic sophistication would not become common in science fiction until perhaps the 1960s or even the 1980s and later.

1947 "E for Effort" T.L. Sherred

It is often argued that a mechanism merely probing and displaying the past in images and maybe sound, while as yet beyond our capacity, is fundamentally a more plausible prospect than time machines reversing into the past. After all, using a time viewer is in essence no more absurd than watching a movie made 50 years ago. You can't change the past events displayed that way, however much you might wish to. Fiction based on a technology of captured images of the past (even a past only a few seconds ago, and perhaps anywhere in the world) has been a minor but significant contributor to the Time Machine literature.

As we shall see in later chapters, this notional device has been a recurrent temptation in sf, with such stories as Isaac Asimov's "The Dead Past" (1956), Bob Shaw's famous "slow glass" tale "Light of Other Days" and its novel-length sequel *Other Days, Other Eyes* (1966, 1972), Damon Knight's "I See You" (1976), and Arthur Clark and Stephen Baxter's title-filching *The Light of*

Other Days (2000).[2] It does not, however, include such ingenious pleasantries as Gregory Benford's "Time Shards" (1979), where ambient sounds from ancient pot makers are decoded from the gravings on their ceramic products, or Sean McMullen's "The Colors of the Masters" (1988), when a lost nineteenth century device is found to have recorded playing by Beethoven, Mozart and Liszt.

Sherred's first story, published in *Astounding* so soon after the end of the second world war and its blazing conclusion in the nuclear bombing of two Japanese cities, captured the paranoiac hysteria of the times (reflected also in the Asimov story cited above) and sent a shockwave through sf readers. It appears as an honored entry in such anthologies as *A Century of Great Short Science Fiction Novels* (1970) and *The Science Fiction Hall of Fame, Volume Two B: The Greatest Science Fiction Novellas of All Time* (1973). Today, in the era of naked self-revelation of Facebook and other social media, personal privacy is already so compromised that readers might not be as moved by time viewer assaults, but there remains in the fiction on this theme a seesaw balance between dystopia (Sherred's despairing story) and utopia (Knight's embrace, and Clarke and Baxter's, of an end to neurotic anxieties of the body and the other).

Ed Lefko is waiting in a seedy part of Detroit on his way to Chicago, and kills some time paying a dime to watch a low budget movie in a store with its windows painted black. The quality of the images is astonishing, the editing and presentation amateur. Ed has a beer with the proprietor and film compiler, Miguel Jose Zapata Laviada, son of Mexicans emigrated to the States in the 1920s. Mike confesses that he created the film, using a device he cobbled together that reaches across duration and locale. Attempts to fund development of this miraculous process has run aground on incredulity and his own reserve. He is ready to sell the secret to Phillips Radio. Lefko takes him in hand, raising an initial tranche by blackmailing several dubious rich men with compromising photos taken using the device. The story roars ahead in caper mode, as Mike and Ed hire a cute blonde to turn away the curious and make an hour of teaser footage of Alexander the Great as a young emerging warrior.

Intoxicated by their promising start and many bottles of champagne, the trio fly to Los Angeles, register at the Commodore, buy fresh clothes suitable for scammers, find Lee Johnson, a "brisk professional type, the high bracket salesman" and introduce themselves as "embryo producers." Johnson has a stable of energetic young movie practitioners and one irritated snarler. The

---

[2] And others earlier and later, listed in the magisterial online third edition of the *Science Fiction Encyclopedia* at http://www.sf-encyclopedia.com/entry/time_viewer

rough cut enthralls them; they set out to finish the movie with unknowns from Central Casting. The finished work is a triumph, and instantly solves their income problem. A second epic, *Rome*, is no less a smash hit, although they are obliged to edit Jesus out after the Catholic and Protestant Board of Review claim their "treatment" is irreverent, indecent, biased and inaccurate "by any Christian standard." Worse, Jesus doesn't look anything like the Hollywood version of the Son of God. History experts weigh in with complaints, and the final cut is trimmed to agree with Gibbon's *Decline and Fall*.

So the story races along, but Sherred was, of course, working toward mass outrage when Ed and Mike finally create the movie they wanted to make, something that will shake the ground under the feet of corrupt politicians, wealthy corporate scoundrels, protected gangsters and rotten cops, and international power dealers with their hands on the new horror weapon, the atom bomb. Planning well in advance, the final movie is run off in numerous copies along with transcripts of the words spoken in the major languages by the most puissant and debased in the world, their time viewer-captured conversations lip-read and synched. Simultaneous premiers open around the world.

The film is denounced as abominable slander, and the actors sought for prosecution or worse; of course there are no actors. Theaters are set ablaze, mobs riot, desperate attempts are made by the authorities to find the perpetrators and seize or destroy their fatally dangerous devices. Ed and Miguel are murdered in their confinement, their last message (the informal history we have just read) found and used to guide officials to the bank vault protecting their blueprints for the time viewer. A nuclear warhead vaporizes the Detroit Savings Bank. We fade to black with the clear implication that spasm nuclear catastrophe has begun.

It is a desperately pessimistic view of human society in the 20th and by implication the twenty-first century, indeed of all centuries. Seventy years after its first publication, "E for Effort" can strike even the jaded and cynical into melancholia as deep as the gloom suffusing H.G. Wells's Time Traveler gazing upon the vile closed ecology of Morlocks and Eloi and then the final ruin of a dying Sun and dead Earth.

1949 "Private Eye" Henry Kuttner and C. L. Moore (as by Lewis Padgett)

Like Sherred's "E for Effort," and in contrast with C.L. Moore's solo "Vintage Season," this viciously sharp psychological drama written with her husband less than a decade before his premature death examines not time *travel* but its cousin, time *viewing*.[3]

If such a technology were not banned outright, or obliterated in an outright nuclear holocaust as in Sherred's pioneering tale, forensic sociologists of the future might use such a device as readily as they now examine DNA samples or spray Luminol to test for latent bloodstain evidence of criminal acts. In such a society, it might not be the *deed* that needs proving, but the *intent* of the perpetrator. Suppose you are badly injured emotionally or financially and harbor a desperate wish for revenge. How can you make your plans and then perform the crime while hiding the slightest glimmer of murderous intent from police viewers in the near future?

The Eye, or tracer, is a time viewer with a 50 year range into the past, but that can be directed by its engineers to any specific target. No criminal act can escape its scrutiny. But in this future, inflicting death can only be punished if intent to murder is provable, often by tracing the killer's actions in the days, weeks or years leading to the deed. Calculated preparations for a murder cannot be hidden. When Sam Clay stabs his employer and benefactor Andrew Vanderman with a letter opener, a repurposed ancient scalpel, the fact is undeniable in a world where a time viewer sees all. There is no privacy from the all-seeing Eye. Sam shows horror and remorse at what he has done, which seems plainly to have been a reflex fear response to a physical assault by Vanderman. The story unfolds as the opposite of a locked-room mystery. The door is wide open, and has been ever since the tracer was invented, and the killing and its antecedents are available as remorseless evidence. "Scop" (the drug scopolamine once supposed to make deliberate deceit impossible) could solve the key question of intent, but in this future self-incrimination by such medical means can be refused, nor may any guilty implication be legally drawn from this choice.

The story is therefore not a classic crime investigation, but like Alfred Bester's subsequent novel of telepathic law enforcement, *The Demolished Man* (1951), it becomes a psychological probe of motive and mental opportunity. In both narratives a kind of psychoanalytic Oedipal conflict is finally uncovered

---

[3] Citations from the Kutter/Moore collection *Two-Handed Engine* (Centipede Press, 2004).

that has driven a man to slay a stand-in father figure. Or perhaps in this case it is childhood trauma of a different kind, a punitive cold father who locked the young Sam Clay in the dark of a closet for his real or imagined infractions, reinforced by a framed picture "of a single huge staring eye floating in space. There was a legend under it. The legend spelled out: THOU GOD SEEST ME" (813). Here is the shaping trauma: parental punishment in the night "for some childish crime," displayed by the time viewer:

> The child was in bed, sitting up wild-eyed, afraid. A man's footsteps sounded on the stair... Moonlight fell upon the wall beyond which the footsteps approached showing how the wall quivered a little to the vibration of the feet, and the Eye in its frame quivered, too. The boy seemed to brace himself. A defiant half-smile showed on his mouth, crooked, unsteady" (814).

That repressed defiance in the face of a father and a punitive God is the key to Sam Clay's terrified determination to have his revenge when handsome, Alpha-male Vanderman steals his girlfriend Bea. To murder his rival without leaving a trail for the tracer to find requires iron control over his actions and reactions, a protected redoubt in his mind that allows him to pretend friendship and gratitude to Vanderman. Step by planned step, he becomes a trusted professional flunky. The narrative allows us to enter his mind, the one redoubt into which the time viewer cannot go. This much can be assumed, Clay tells himself in that sealed sanctum: "If I stand to lose by Vanderman's death instead of gaining, that will help considerably... I must make it seem as though he's done me a favor—somehow" (796–97). Bea now sees him "as an exaggerated symbol of both romance and masculine submissiveness" (805).

Details of his ingenious vengeance must be reserved for the reader, but it proves sufficient to let him escape the accusing gaze of the Eye, declared innocent of murder. But this success leaves Sam bereft of a goal. He no longer desires Bea, who attempts to manipulate him, contemptuously insisting that he was "never a planner." He laughs uproariously at this, embarrassing Bea, but understands finally what the time scanner could never quite delve deep enough to find. "Killing Vanderman hadn't been the answer at all. He wasn't a success. He was a second-rater, a passive, helpless worm... I hung dark glasses on an Eye, because I was afraid of it. But—that wasn't defiance" (819). Clay smiles crookedly, glancing from Bea to the ceiling.

> "Take a good look," he said to the Eye as he smashed her skull with the decanter. (819).

The notable sf writer and sometime-critic Barry Malzberg suggests that this disturbing Kuttner *Astounding* story might indeed owe a debt to Sherred's "E for Effort" (1947) which was published in the same issue as the first part of their own post-nuclear holocaust serial, *Fury*. It uses the time viewer "in a close-in, hermetic context to deepen and perhaps extend Sherred's simpler point. Sherred saw technology's promised marvels leading inevitably to societal death and disaster (because once a device, no matter how dangerous, is invented, it will be deployed)." Henry Kuttner, possibly writing alone for this story, "makes the tragedy personal, shrinks it to the level of terrifying intimacy. I think it is one of the greatest of all post-Tremaine sf novelettes. Top ten."[4] (F. Orlan Tremaine preceded John Campbell as editor for 50 issues of *Astounding*, and is notable for encouraging "thought-variant" plots that shifted sf away from pulp adventure into the exploration of new ideas—which is clearly the domain of these related minatory stories by Sherred and "Padgett."

For all its power, they argue, the time viewer cannot ameliorate the resentment, lust, greed and crushing sense of inadequacy or cruel hunger for power that distinguishes *Homo sapiens* from the world's other species. Advances in technology can only multiply the dire effects of these failings. It was an odd perspective to find in Campbell's technocratic magazine, and curiously revealing that stories casting such a surprising shadow should have been instantly applauded, marked for admiration in the decades that followed. This ambivalence is intrinsic to a mode of fiction that can forecast horrific atomic weapons as well as voyages to the stars, cures for disease, even personal immortality. Perhaps it is a consequence of our evolved bent that we are more addicted to tales and images of violent contests than to optimistic visions of peace and flourishing. Even so, as we shall see, the imaginary tools of time viewers and time transports provide as well the basis for hope, however constrained by humanity's current flaws.

1950 "Flight to Forever" Poul Anderson

A steady and voluminous writer of sf and fantasy, Poul Anderson (1926–2000) was not yet 24 when this rousing tale of one-way time travel to the future was published in a less than elegant magazine rejoicing in the title *Super Science Stories*. Nearly three decades later, he was appointed a Grand Master by the Science Fiction and Fantasy Writers of America (but actually of the world), which arguably he ought to have been given years earlier. "Flight to Forever"

---

[4] Personal email from Mr. Malzberg, November 15, 2018, cited here with permission.

is by some measures a crudely wrought adventure story with an impressive but entirely implausible denouement, yet it is not without poignancy and a shivery sense that, yes, this could be what a journey to the end of time, and beyond, might be like for the earliest time traveler in a built-at-home time machine.

The time projector device itself is more persuasive, somehow, than the Wellsian variety, maybe because it has an engine using fuel readily available in 1973, its year of departure for the future. To readers today, Anderson's 1950 depiction of 1973 is technologically retarded, with vacuum tubes instead of transistors, but it feels convincing enough. In scientist Martin Saunders' "capacious underground workshop," the projector

> stood in a clatter of apparatus under the white radiance of fluoro-tubes. It was unimpressive from the outside, a metal cylinder some ten feet high and thirty feet long with the unfinished look of all experimental setups. The outer shell was simply protection for the battery banks and the massive dimensional projector within. A tiny space in the forward end was left for the two men. (208–09)[5]

The second man, Sam Hull, is a loud, large, buoyant engineer. A tech, small and owlish MacPherson, gets a hand wave but plays no further part. The final significant figure is Martin's fiancé (every sf tale of that vintage or earlier required one), the sweet and kind and lovely Eve Lang. Her name might excite a qualm in some readers, but luckily she is not finally marooned in remote history with a man named Adam, as happens in a van Vogt story, "Ship of Darkness," published two years earlier and redeemed by Van's patented oneiric touch.

Following Wells's lead, the explorers have already sent a pair of test instruments a century into futurity, and themselves gone back in time to 1953 and forward to 1993, without suffering any difficulties. True, the two probes failed to return, but they dismiss this as due to the frailty of tubes that "blew their silly heads off"; prepared this time, their craft carries plenty of spare and suitable equipment. They depart. Time travel is noisy in this universe: "the drone of the projectors filled the machine with an enormous song"; when Saunders slaps a switch, "the noise and vibration came to a ringing halt" (211). The house has disappeared. The time machine lies in a half-filled pit. There is no sign of the automatic probes, but at least the ground-level detector "automatically materialized it on the exact surface" (212). How could this work? We are not told; even from the swirling grayness of intertemporality some kind of

---

[5] Citations from the Anderson collection *Past Times* (Tor, 1984) and to Anderson's work throughout granted by The Trigonier Trust.

radar-like signal must be emitted and acted on instantaneously. An even more remarkable trick prevents

> disastrous materialization inside something solid; mass-sensitive circuits prevented the machine from halting whenever solid material occupied its own space. Liquid or gas molecules could get out of the way fast enough (212)

—although not, as Martin will learn later, to his horror, if the machine is trapped under an oceans-worth of water. For now, they head for home, stopping now and then to search for the lost automatics. When they find them, the energy is entirely drained from the probe engines. Martin speculates that "the farther back we go, the more we use per year. It seems to be some higher-order exponential function" (215). What to do? They can never return, because something like limiting velocity of light in vacuum must be operating in their spacetime trajectory. Their only hope is to travel forward in the expectation of scientific advances that will solve this Einsteinian restriction.

Soon the narrative veers from *Popular Mechanics* to *Men's Adventure* mode, when they emerge in 2500 and come swiftly under attack by the Armageddonists, or Fanatics, a blend of Nazis and jet-propelled Inquisitors, who instantly strafe Sam Hull and blow him apart. Nearly paralyzed by grief, Martin Saunders reaches his machine just in time. "Thank God the tubes were still warm!" (221).

In the year 3000, he is pursued by a military patrol, and taken in hand by a street-wise weasel named Belgotai of Syrtis (so we know he is from Mars), who shouts him a tumbler of rotgut, works out that Saunders is from the past, and together they escape from this dreary down-at-heel epoch to the next century, which is a waste land of fused, radioactive rock. In 4100, they emerge on a grassy sward where young men and women in science fictional tunics and capes find them fascinating and bring out a device that receives emissions from their cerebral speech centers, allowing them to follow the conversation and be understood in turn. This is a tool Wells's Time Traveler could have used to advantage, although it is possible that the Eloi were too intellectually regressed and the Morlocks too beastly for this to be very rewarding. Again, the news is bad. A faster than light drive has been known for half a millennium, "warping through higher dimensions," but infinite energy is still needed to travel in reverse for more than 70 years.

They go forward, and find themselves trapped. In 25,296 they are freed from an immense tetrahedral pyramid half a mile high, which has been slowly falling apart for eons. Visiting scholars from space inform them that this monstrous edifice is a mark of the vanished alien Ixchulhi. Now there is a Galactic

Empire, and Earth is of little interest. This 10,000 year old polity is "peaceful, prosperous, colorful with a diversity of races and cultures". Yes, there is a rabble of barbarians along the outer edge of the galaxy, but they are merely a nuisance. It can be proved beyond doubt by new mathematical techniques. Inevitably, though, things are worse as they proceed into futurity. In 31,000, civilization returns, ruled by the Matriarchy. The two time travelers are asked firmly if politely to be on their way.

Things get worse again, the entropy of history. In 50,000, there is another immense construction in a climate of rain, snow, ice, raw wind. A friendly alien and Vargor, an aristocratic warrior, usher Saunders and Belgotai to the drafty fortress, the hold of Brontothor, final redoubt of the Empire and dwelling place of the unbelievably beautiful, serene and really hot Empress Taury. One has the undeniable suspicion that it's not only the empire that's degenerating—for example, they have forgotten how to travel in time—but also the narrative. The mighty galactic fortress of… *Brontosaur*, with a lisp? (When I was a kid first enthralled by sf, I loved this story, and would have dismissed such captious complaining with a curl of my lip. Decades later, alas, one feels the need for a tumbler of rotgut from the cellars of 3000AD.)

The barbarians are coming, their fleet greatly outstripping the handful of Imperial craft. Saunders comes to the rescue with plans for converting their one Super into an armed time machine. With this pivotal advantage they destroy the enemy's own dreadnought and send the rest fleeing. Taury is grateful, and admits that she is "so lonely" and has fallen in love with Martin. Prince Vargor is not pleased. Saunders awakens in restraints inside the time machine, and is sent into the next step of his non-return ordeal. By the time he has get free of his bonds, it is 10,000 years later and he is sunk deep into a new ocean. In the four millionth year he emerges into a city of unbearable puissant energies, and a Voice who speaks all in caps instructs him to flee before the forces will destroy him. "YOU MUST GO ON TO THE VERY END OF THE UNIVERSE, AND BEYOND THE END" (281). Luckily the super posthuman does a refit of the machine, providing a new power source and presumably food and water for the all-but-infinite journey.

Now an aspect of authentic majesty suffuses the closing pages. Ever more swiftly, into the darkening universe of dying stars and ever outwardly-rushing ruined galaxies, Saunders is carried for billions upon billions of years. Sunk in apathy, he sees in the intolerable blackness of the dead cosmos a glimmer of faint light. "*The universe was reforming*" (285).

There has been no Big Crunch back to a Big Bang singularity, just the re-emergence of a Laplacean universe where every particle somehow recovers the memory of its earliest previous state and position and quantum energy (but presumably without the unpredictable probabilistic jitter of Heisenberg uncertainty to disrupt this wholesale *Risorgimento*). And yes, here is a boiling world congealing out of its molten birth, and rain, budding ferns and mosses, glaciers, animals; the Moon, with its scar-pocked face, is exactly as he recalls it. The Einstein continuum "was spherical in all four dimensions… if you traveled long enough, through space or time, you got back to your starting point" (287).

At last Martin Saunders is home, just after he left. The machine dissolves behind him, as the gods have ordained. He could reconstruct it, but he will not do that; he will not permit time travel's opportunity for oppression and murder. (He does not reflect, though, that this recycling of history means that many others after him are pre-determined to invent time travel for themselves, and then to lose it in the entropy tides.) He has lost Taury, but his kind and loving Eve awaits him. "It was enough for a mortal man" (288).

6

Empires of Time

1953 *Bring the Jubilee* Ward Moore

One of the most persistent themes in science fiction and philosophical reflection alike is the question "What If—?" Give me a time machine, as Archimedes did not say, and I will change the past. But is it genuinely possible to modify what is already part of the historical record, even with time travel? Or is the logical impossibility of rewriting the known past sufficient evidence that time machines just cannot exist, no matter how much power and subtle knowledge is available?

Physicists and philosophers have argued in favor of strict limits to reverse causation, as we have seen, arguing at best for a restricted ability to contribute to past events only if the outcome abides by the need for overall consistency. What you *can't* do, various theorems insist, is twofold: (1) prevent your own conception (which would create an unphysical and self-defeatingly paradoxical time loop), or even (2) alter any circumstantial condition in the past that must lead to a road block in established history, or its subversion by means as yet unknown.

Well, these objections to time machines seem rather plausible, especially the first. But what might "circumstantial" entail? Throttling Genghis Kahn or Hitler in the crib? Or merely accidently crushing a butterfly in the remote past, as Ray Bradbury's short story speculated? How would that speculation even be tested, since per hypothesis the pre-modified history must be knowable only to the perpetrator or perhaps not even then. The science fiction ironic humorist William Tenn—*nom de plume* of Philip Klass (1920–2010)—traced this process through a series of iterations in "Brooklyn Project" (1948),

© Springer Nature Switzerland AG 2019
D. Broderick, *The Time Machine Hypothesis*, Science and Fiction,
https://doi.org/10.1007/978-3-030-16178-1_6

a time machine story deliberately echoing the nuclear weaponry Manhattan Project. Driven by environmental and evolutionary factors, consecutive turns of the cycle result in ever-more zany alterations, none of them noticed by the increasingly morphed military and government officials:

> ...their bloated purpled bodies dissolved into liquid and flowed up and around to the apparatus. When they reached its four square blocks, now no longer shrilling mechanically, they rose, solidified, and regained their slime-washed forms.
>
> "See," cried the thing that had been the acting secretary to the executive assistant on press relations. "See, no matter how subtly! Those who billow were wrong: we haven't changed." He extended fifteen purple blobs triumphantly. "Nothing has changed!"

At the core of *Bring the Jubilee* by Ward Moore (1903–1978) is a version of this time travel outcome, which is not so much "*What If* the Confederates had won the Civil War and someone sets out to alter that history?" as "How could one time traveler bring about such an enormous change, and what would the process look like?" Hodgins (Hodge) McCormick Backmaker was born in 1887 in the run-down northern New York town of Wappinger Falls, and by the age of 17 was a tall, strong youth with physical endurance and a hunger for books but lacking useful manual dexterity. His father is mild, his mother a tough-minded foe of "idleness and self-indulgence."

Hodge's dreams of an education at Harvard or Yale seem hopeless, but he sets off by foot on the 80 mile trudge to New York city and once there swiftly ends up coshed by a robber in a dark alley and without his shirt, shoes or minimal savings. It is his dubious fortune to be taken in charge by an alcoholic fellow, George Pondible, who helps him steal shirt and shoes from an even more derelict wretch in the alley and gets him a drink and meal. Pondible introduces him to a nihilistic trader in whose book, printing and stationary store Hodge remains for the next six years, reading voraciously.

As we look over his shoulder, we imbibe a great deal of backstory about this alternative history, in which slavery has been abolished by "chivalrous Southrons" but indenture is widespread. With the North humbled and restricted, advances in technology are seriously stifled (gas lighting is common, no electricity, telephony and heavier-than-air flight are dismissed as impossible but advances have been made in balloon flight). When Hodge finally leaves the bookshop after an invitation from the sexually electrifying young physicist Barbara Haggerwells to study among the independent scholars at Haggershaven, his devotion to history is encouraged. In due course he develops a certain reputation even among the professors of the great universities of Germany and the South.

Barbara is a neurotic and frequently hostile termagant who appears driven by rage against her dead mother. A proto-psychoanalyst comments on her plight:

> Her fantasy of going back to childhood (fascinating; adult employs infantile time-sequence, infantile magic, infantile hatreds) in order to injure her mother is a sick notion she cherishes the way a dog licks a wound. (108)

Still, Barbara proves sufficiently brilliant that she replicates crucial work on spacetime that in our history would be Einstein's great achievement. Propelled by this bitter yearning to punish her mother, she invents a time machine and persuades the assembled scholars and artificers of the gentle colony to help her build it in a disused barn. Trials are conducted; the machine works. Hodge is chosen to voyage back to 1863, where he will have the unprecedented opportunity to witness the pivotal battle at Gettysburg and study precisely how the South bested the troops of the North.

It is only at this late point in Hodge's journal, written in 1877 and which we have been reading, that he ventures upon the historical crux of American self-definition, and of the novel's tangled threads (most of them scamped here). He will return to the location of the barn-laboratory in time to be recovered at midnight on July the Fourth, 1863. As the device activates, he experiences

> the shattering feeling of transition. My bones seemed to fly from each other; every cell in my body exploded to the ends of space.... Most of all, I knew the awful sensation of being, for that tiny fragment of time, not Hodgins McCormick Backmaker, but part of an *I* in which the I that was me merged all identity. (171)

The unpleasant shock of time travel is brief. He heads for the road to Gettysburg, some 30 miles distant, and walks there cautiously but unscathed. "On my left I knew there were Union forces concealed; on my right the Southron maneuvered. In a few hours to walk between the lines would mean instant death, but now the declaration had not been made" (177). He waits in an orchard, relaxed in grass as a cavalry in blue moves through. The Confederates follow, some clad in the semi-official butternut yellow, streaked and muddy but clearly representing "a victorious, invading army."

Of course his much cleaner apparel attracts the interest of several Confederate soldiers; one demands his boots. Hodge refuses, declaring his civilian status. A tall captain with distinctive features interrupts the theft. The man is not a notable figure from the history records, yet Hodge is convinced he recognizes the face. Abruptly, with no deliberate history-changing intervention by Hodge, this minor contretemps turns into a frantic retreat from Round Top as Confederate soldiers conclude that Hodge is a scout from "the blue-bellies" who are "laying' fur us." The captain fails to halt the hysteria, seizes one of the soldiers, and in an accident the man's musket discharges and blows the captain's face away. Hodge fumbles for the corpse's papers, but is too sickened and remorseful to read his identity.

He sees the murderous exchanges at Gettysburg, with small alterations on the first day due to the interruption in troop positions that his presence has precipitated. On the third day, because this time the Federals held the Round Top, Pickett's famous charge fails, becoming "a futile attempt to storm superior positions... ending in slaughter and defeat" (183). Hodge staggers away, finally, and reaches the barn with two hours to spare before the time machine is to draw him back to the future. To his horror, this does not happen.

At last he identifies the dead Captain, recalling a prominent portrait hung in the library at Haggershaven: its founder, "Herbert Haggerwells... never to become a major now, or buy this farm. Never to marry a local girl or beget Barbara's great-grandmother. Haggershaven had ceased to exist in the future" (185). So the time loop, not a traditional paradox but rather time shunting from one track to another, leaves Hodge at the dawn of our own age, a world with horrors of its own pending in its future and with technological wonders unimagined in Haggershaven—but with no time machine. Not yet, anyway.

1954; 1973 "Beep"/*The Quincunx of Time* James Blish

If Ward Moore's *Bring the Jubilee* is an emotionally complicated "What If—?" novel, the 14,000 word story "Beep," written in the same year by James Blish (1921–1975) and expanded 20 years later as *The Quincunx of Time*, is even more conceptually complicated but emotionally torpid. We shall look at the Quincunx version, which includes additional speculative material; "The drama, for those capable of enjoying it in this form," Blish notes in a "Critical Preface," "lies more in the speculations than in the action... (1)." It is an exercise in presenting an idea briefly and densely, almost a Socratic dialogue (if Plato had been a detective trained in quantum theory) told by actors in funny hats—although wardrobe changes and disguises are a key if ludicrous element.

It, like the Moore book, investigates What If a Time Traveler Intentionally or Accidentally Changes the Past, not with a Wellsian time machine or even Barbara Haggerwells's projector, but via messages intercepted from the future. (That remark was a Spoiler of interstellar proportions; sorry, but think of it as an item of information provided to you by Blish's Dirac transmitter.)

Here is the set-up: a bland and bored field agent with the Service, code-named Jo Farber, sits on a park bench on the planet Randolph under twin suns and keeps an eye on a young man and woman who meet, doing so safely and without interruption. Farber has worked out that the Service is safeguarding the future children of this couple, which means that it knew not only when the couple are destined to meet but also who their unconceived children will be and what significant role will come their way in future decades. It should be obvious from the news reports, he reflects, that the Service has some arcane technique for accurately predicting the future. Major attempts at military and other attacks on the status quo are met in the nick of time by superior force maximally emplaced to win. He removes his glued-on disguise and put his deductions to his Service boss, Krasna, who promotes him on the spot and shows him a dramatization of events several hundred years earlier.

A beautiful and blazingly intelligent media journalist, Dana Lje (this Serbian-Cyrillic ligature would be a handy clue if it were pronounced "lie," but it's more like "ll") brings Captain Robin Weinbaum a list of predictions she claims have been passed along to her by a source calling itself Interstellar Information. These prove so accurate, impossibly so, that Robin threatens her with jail as a major criminal or an accomplice, and for violating the Official Secrets Act by simply knowing about an invention called the Dirac communicator. Dr. Thor Wald is fetched in, as the inventor of this secret and classified device; he is very tall, blond, gorgeous and seems harmless except to vulnerable women, which is all of them. He explains the breakthrough occasioned by this instant faster than light messaging machine based on a conjecture of the Nobelist Dirac at the dawn of the quantum theory age. Previously, only the ultrawave has beaten light-speed by riding on a phase velocity transmission (if you're a physicist you will know about this; if not, don't ask) and not by very much.

Let's pause for a moment. With a degree in biology and a compulsive interest in other sciences as well, not to mention being a minor authority on James Joyce's opaque masterpiece *Finnegans Wake* and music both classical and experimental, Blish was exemplary for his convincing sf doubletalk often based at least loosely and provocatively on authentic science (or exploratory varieties like British astrophysicist Edward Milne's revision of relativity theory, now long superseded). When Blish wrote *Jack of Eagles* (1952), a novel based on psi phenomena, a fad in early 1950s' sf, he did not just handwave; he offered equations by Baron Blackett (1897–1974) and an elaborate model of multiple universes before Hugh Everett (1930–1982) introduced his Many Worlds model in 1957, now the darling of cosmologists.

The prorogation velocity of a Dirac pulse, according to Dr. Wald—"which is not an energy transfer system at all in the usual sense"—is effectively instantaneous. There's just one drawback in this system; incoming messages are invariably preceded by a brief *beep* of meaningless noise. When the device is used for operational purposes, this irritating artefact is trimmed off and deleted. It turns out, eventually, that the beep comprises a massively squeezed version of every single Dirac message ever sent, and *that will ever* be *sent*.

How can these superposed signals be unpacked and separated, then decoded, so Robin Weinbaum's Service team can use the beep to track future felons? The answer is to "tune individual messages out of the beep by time-lag, nothing more… [Because] there are relay and switching delays, the beep arrives at the output end as a complex pulse which has been 'splattered' along the time axis for a full second or more" (94). This can be accentuated by slowing its recoding, exaggerating some effects and diminishing the rest, and suppressing background noise. The result can be crystal clarity. Wald observes that this Dirac effect "poses a metaphysical problem [in regard to causality and choice] quite staggering in its implications" (96).

What evidence is there for this entire bizarre and unprecedented signaling method? Dana cites "the commander of a world-line cruiser, travelling from 8873 to 8704 along the world-line of the planet Hathshepa, which circles a star on the rim of NGC 4725, call for help across eleven million lightyears—but what kind of help he was calling for, or will be calling for, is beyond my comprehension" (96–7). Reading this at age 14 or so, more than half a century ago, I found it wondrous in the extreme. Despite a strong suspicion these days that Blish was flummoxing me at every turn I still find cognitive breakthroughs or paradigm breaches enthralling. Less so is some of the rest of the plot, such as it is. It turns out that the old man who represents himself to Captain Weinbaum as the original inventor of the Dirac device, one "J. Shelby Stevens," is really Dana Lje in persuasive makeup and costume, adopted to protect her from arrest before she has proved the worth of decoding the compressed messages. Far easier to believe the theory of Dirac transmission than that comic opera silliness.

Does this delightful conceit get us and sf any closer to a rational account of time travel, and the difficulties of free will in a universe where beeps provide exact advanced report of future choices? Two centuries after Dana's exploits, Krasna tells his new recruit that

From the four dimensional point of view which we provisionally adopt—and please note well that 'provisionally'—we know that the consciousness of the observer is the only free thing in the universe.... Our obligation as Event Police is to make the events of the future possible, because those events are crucial to the evolution of our society... (125)

Does that really make any sense? Not to many philosophers of science and mind, I think. But if time travel of this non-corporeal kind if even developed, we will have an empirical answer—and arguably that is the soundest basis for claims to theorized knowledge, once the hard work has been done.

1955 *The End of Eternity* Isaac Asimov

The cosmos of Blish's "Beep"/*Quincunx* can be construed as a block universe, where the future is always-already written, while Ward Moore's world is so fragile in time that a single traveler to its past can unhinge and divert history. Asimov's best novel, *The End of Eternity*, postulates a past, present and future of endless variation, manipulated by "Eternals" who dwell outside ordinary duration and ceaselessly manipulate history on a massively industrial scale, making the most minor of calculated changes.

One of their key goals is to obliterate any future that leads to space travel and the rise of an interstellar empire precisely of the kind detailed in Asimov's *Foundation* sequence of novels—the originary source of this Russian-American polymath's reputation as a science fiction genius. Given the novel's title, and Asimov's fondness for his futurist spin on Edward Gibbon's *Decline and Fall of the Roman Empire*, we can be confident that the Foundation history will survive and Eternity's will end, but the satisfaction is in the details and the wily path through time loops and paradoxes to that foretold conclusion.

In literary terms—the way it is written, the specificity and development of its characters, the robustness or delicacy of its *mise en scène*, and so on—there is plenty wrong with the novel. We can be justifiably less concerned with such artistic critiques, in this book, and instead apply sharper focus to our chosen topic: the imagined operating principles and consequences of time travel. By "operating principles" I do not mean just the physics of time hopping to and fro through years, centuries or millennia, but the economic and communal factors that keep it running and shape its organization arrangements.

Other novels or sequences of the 1950s drew upon military and political chains of command (Poul Anderson's series about a Time Patrol keeping history safe from partisan rewriting, for example, or H. Beam Piper's and Keith

Laumer's somewhat related crosstime policing of alternative histories or contests between them). Asimov's version, Eternity, seems based on the pocket universe of academia: the university, rather than the nation state or empire or capitalism versus communism. Eternity is an immense bureaucracy in a constructed and literally artificial reality parallel to Time. It stretches in its infrastructure command upwhen and downwhen from our current millennium (aside from a mysterious blank period blocked from access) to "Nova Sol" billions of years hence when the Sun explodes and provides inconceivable amounts of free energy ready to be tapped by the Solar System's past.

Who runs this enormous institution? Why, an unmarried meritocratic clergy somewhat like ecclesiastical scholars: Sociologists, Educators, Observers, the administrators of the Allwhen Council, Life-Plotters, analysts such as Senior Computers (where a "Computer" is a professional evaluating possible future Reality Changes, rather than a machine such as an artificial intelligence) and Technicians who do the dirty work—but not the lowly Maintenance crews who keep Eternity ticking over.

At 15, Andrew Harlan was inducted from the ninety-fifth century into Eternity, where he was schooled for a decade as a Cub before becoming an Observer and finally a Specialist. His ambition was Computer but he was made a Technician, entering briefly into Reality and making some small change that will shift future history into a safer path less likely to collapse into catastrophe. One downside to this custodianship of Reality is the repeated blocking of space flight, seen as a wasteful diversion. Another was the frequent extinction or modification of millions of existing individuals, a kind of soft genocide that made Technicians abominable to the rest of the Eternals, displacing and cleansing their own guilty participation. At 23, he is embittered by what he sees as his exile to purdah, and ready to tilt into actual criminal betrayal of his code when he meets the ravishingly lovely Noÿs Lambent, visiting aristocrat from the decadent four hundred and eighty-second century. She subtly expresses interest; they make love; he learns that her century is due to be renovated, perhaps extinguishing her existence. Harlan cannot allow this, and spirits her away far upwhen to safety. No paradox is risked, as she appears to have no descendants or indeed ancestors. It is a weakness of Asimov's plotting that this remarkable detail would not have set off any alarms in the ace Eternal's mind, but then he was besotted.

Meanwhile, Harlan has been put in charge of the special training of a Cub named Brinsley Sheridan Cooper of the 78th. Cooper proves to be critical in solving a persistent mystery of Eternity: where did this officious parallel timeline come from? The key ingredient of time travel was the mathematics of the Temporal Field, created in the twenty-fourth century by Vikkor Mallansohn but not used for building a time machine for another three centuries when research by Ian Verdeer in the twenty-seventh century provided the basis for the fundamental Lefebvre equations. These in turn made Eternity possible, since as Harlan explains "Eternity is only one tremendous Temporal Field short-circuiting ordinary Time and free of the limitations of ordinary Time." Cooper will be trained in these arcana and sent back to the Primitive history before the invention of time travel, ready to introduce this insights to Mallansohn.

But Mallansohn dies in an accident, so Cooper takes over his identity, lives out his life in the 24th and writes a memoir that specifies the roles of Harlan and Senior Computer Twissell, Andrew Harlan's aged sponsor and superior. Here is a time loop of apparent paradoxical impossibility, yet it is self-sustaining unless Harlan breaks it. In league with Noÿs he does this, although he is bitterly convinced that she has been manipulating him, and is probably an agent from the hidden years beyond the year 100,000. She admits it, explaining that the very existence of Eternity leads to the destruction of humanity. Without hyperspace space travel to the stars, competing alien civilizations will thrive and leave humankind stranded in a backwater of galactic insignificance. The future world will be left a radioactive ruin. Harlan is persuaded, and obliterates Eternity before it begins, freeing up the chance for "the beginning of Infinity."

Praised on publication for its ingenuity, *The End of Eternity* remains a significant contribution to science fiction's explorations of time travel. The actual time machines are devices called "kettles," perhaps for a fancied resemblance of shape. "Its sides were perfectly round," we are told in the opening paragraph, "and it fit snugly inside a vertical shaft composed of widely spaced rods that shimmered into an unseeable haze six feet above Harlan's head" (7) It does not move in space. "The kettle he left, of course, was not the same as the one he had boarded, in the sense that it was not composed of the same atoms. He did not worry about that any more than any Eternal would. To concern oneself with the *mystique* of Time-travel, rather than with the simple fact of it, was the mark of the Cub and newcomer to Eternity....He paused again at the infinitely thin curtain of non-Space and non-Time which separated him from Eternity in one way and from ordinary Time in another" (8).

While that does not make a great deal of sense, it is not less convincing than Wells's time machine with its bicycle seat. It is sometimes complained that Andrew Harlan is a sour, resentful, pitiful protagonist and that this makes the book unreadable. That is an unsophisticated and limiting way to understand how fiction functions, so we can dismiss it.

However, the psychology creaks badly. Throughout, Noÿs is portrayed as a sort of girly-girl, not so much seductive as compliant, expressing surprise and astonishment at facts she knows and that form the basis of her mission. Once Harlan announces his deduction that she is a manipulative agent from the Hidden Centuries, she instantly hardens and gains about 30 IQ points before explaining that she has been in love with him for years before they met. And so on. In short, the emotional underpinnings of the love story are bogus, necessary mostly to prepare the rabbit so it can be pulled from the hat.

Is the argument against Eternity and time travel obviously in error? Senior Computer Twissell presents it:

> *You* know Primitive history, Harlan. You know what it was like. Its Reality flowed blindly along the line of maximum probability. If that maximum probability involved a pandemic, or ten Centuries of slave economy, a breakdown of technology, or even a—a—let's see, what's really bad—even an atomic war if one had been possible then, why, by Time, it *happened.* There was nothing to stop it.
>
> But where Eternity exists, that's been stopped. (159)

Noÿs offers the counter-argument:

> Man would not be a world but a million worlds, a billion worlds. We would have the infinite in our grasp. Each world would have its own stretch of the Centuries, each its own values, a chance to seek happiness after ways of its own in an environment of its own. There are many happinesses, many goods, infinite variety…. *That* is the Basic State of mankind…. (185)

Perhaps so. Yet must time travel necessarily lead to a paralyzing technocratic *faux*-benignity? A similar fear was expressed several years earlier in Jack Williamson's *The Humanoids*: small hive-mind-connected robots swarmed over the Earth, preventing danger and injury and crushing the human spirit. Asimov's own robots might be imagined as having the same dire proclivities, and indeed in the Foundation trilogy that continued Asimov's own sequence we learn that robots from Earth quietly exterminated all the alien intelligences in the galaxy, the kind of truly disgusting "cleansing" genocide that the philosopher-kings of Eternity practiced on the humans of Time.

1955—*Time Patrol* Sequence Poul Anderson

In the same year Isaac Asimov's *The End of Eternity* was published, so was the first story, "Time Patrol," in what would become an influential series by Poul Anderson (1926–2001). As in Asimov's novel, the time travelers influence history, but almost always to protect rather than modify its shape, or to reverse injuries done to it. Because Anderson's temporal police largely dealt with events in the known past, although often manipulated by rogue forces from the future, the concatenated sequence created over four decades was far denser than Asimov's, richer with sensory impressions and complex motivations. These novellas and one novel were gathered in two hefty volumes, *The Time Patrol* (1991, but including stories previously gathered in *Guardians of Time*, 1981) and its sequel, *The Shield of Time* (1990), plus one short story released six years before Anderson's death. The cumulative effect is impressive, as is the conceptual clarity undergirding the saga.

The narrative mostly follows a highly competent, thoughtful white American, Manse Everard, recruited in 1954 to the Time Patrol at the age of thirty after wartime service and subsequent adventures. His training for this unusual profession was conducted using advanced pedagogical techniques at the Academy—located in the American West in the Oligocene period, built a thousand years earlier, destined to be maintained for another half million before scrupulous demolition to leave no traces for archaeologists to stumble upon. Transit in time would not be discovered until the year 19,352, leading to the retrospective creation of the Patrol and establishment of the Academy. Its deep motivation was the desire of a posthuman species, the all-but-omnipotent Danellians, to safeguard their own temporal trajectory from earliest *Homo sapiens* to their emergence more than a million years from now.

The Patrol maintains offices in many centuries, warehousing local garments, language imbuers, and hoppers or timecycles, scooters that somewhat resemble two-seater wheel-less motorbikes but fitted with antigravity, invisible protective force shields and the necessary time travel equipment. Everard soon attains the rank of Unattached, permitted to roam in any era and resolve problems with the aid of information recorded in the future but carefully screened to avoid time loops. Anderson's universe in this sequence is mutable; in one story, the entire Patrol structure is abolished, leaving a handful of operatives to undo the damage and return it to reality. So knowing what the records say about your current task and its successful resolution (or otherwise) offers no shortcuts to hands-on research and action. If history is changed, so too will be the records—but only after you have brought about those alterations.

On the other hand, mutability is not inevitably effective in enforcing change over time. We learn that "it's rather as if the continuum were a mesh of tough rubber bands. It isn't easy to distort it; the tendency is always for it to snap back to its, uh, 'former' shape. One individual insectivore doesn't matter, it's the total genetic pool of their species that led to man" (10). Kill a single living creature and all that's lost in the long term is that particular arrangement of widespread genes that persist, subject to the same selection pressures and opportunities in a restlessly changing environment. Of course, in human terms the loss of a single person or group can make an enormous difference, and so the Patrol is needed for stability in the flight of the existing arrow of history.

Astute and decisive, but also empathic, often grieving at the losses occasioned by his logical interventions, Everard deals with stresses and disruptions in the lanes of time. A lost patrolman is obliged to take over the role of the Persian Great King Cyrus and then abandon it to return from his magnificent Persian consort to his thin-voiced wife in the present. A manipulated history makes Carthage the victor over Rome, and all of subsequent history is rewritten, until Everard and a Venusian from his own future set things right. He deals with trouble among Goths, and dangerous Exultationist warriors to the Peruvian city Caxamalca under Incan emperor Atahualpa. All of this, and much more, is great fun, but for our purposes it is the nature of time and command over it that is especially interesting.

As with all time fiction, most of the technical explanation is pretty much pure handwaving. The form of math and science is there, but the content remains to be filled. In the opening stanza of *The Time Patrol*, Everard and other inductees are told that travel into the past "requires infinitely discontinuous functions... it involves the concept of infinite-valued relationships in a continuum of 4N dimensions, where N is the total number of particles in a universe." What's more, "time is a variable; the past can be changed." A young physicist objects that what is implied is logically impossible, because an event "cannot both *have* happened and *not* happened That's self-contradictory."

Not so: "Only if you insist on a logic which is not Aleph-sub-Aleph-valued" (6). Go back and prevent your parents from meeting, and while that portion of history would "read differently; it would always have been different." Still, because you belong to a section of history prior to the change you induced, you must remain alive and participating in the new reality. Objecting that you cannot exist without an origin in the renovated universe is to insist, incorrectly, that "the causal law, or strictly speaking the conversion-of-energy law, involve only continuous functions. Actually, discontinuity is entirely possible" (7). And "when the past has been deformed, the Patrol does annul the events that flow from it" (457).

In the closing pages of *The Shield of Time*, a Danellian speaks:

> "Think… of diffraction, waves reinforcing here and cancelling there to make rainbow rings. It is incessant, but normally on the human level it is imperceptible…. In a reality ever liable to chaos, the Patrol is the stabilizing element, holding time to a single course. Perhaps it is not the best course, but we are no gods to impose anything different when we know that it does at last take us beyond what our animal selves could have imagined. In truth, left untended, events would inevitably move toward the worse. A cosmos of random change must be senseless, ultimately self-destructive. In it could be no freedom.
>
> Has the universe therefore brought forth sentience, in order to protect and give purpose to its own existence? That is not an answerable question.
>
> But take heart. Reality *is*. You are among those who guard it. (435)

This might strike us as rather a self-serving doctrine from one of the near-gods whose existence depends utterly on the preservation of an ordered world when alternatives or "deviations" are annulled at their will. What they are imposing on history is a kind of converse of the labors of the Eternals, who strive to bend the arc of time away from a galactic imperial future toward their preferred safe, conservative goal—one which, according to the Asimovian version of Danellians, will lead to enfeeblement and the extinction of the species. So both these narratives of a world sanctioned by superbeings from the remote future propose the same sort of rationale. The difference, if I've understood this correctly, is that Asimov's superbeings, all but Noÿs herself, will be obliterated retrospectively in the same instant that Eternity is negated, has never been. On the other hand, we are told that her people in the Hidden Centuries discovered a form of time travel of their own, and that the future is an infinite sheaf of possible outcomes, so perhaps they diverge into persistence.

Such is the bafflement of the Time Machine hypothesis; we probably need an Aleph-sub-Aleph-valued logic to reach a convincing answer, but sadly, just at the moment, it remains out of our reach.

1956 *The Door Into Summer* Robert Heinlein

One obvious means for traveling into the future that we haven't discussed seems to be a cheat when the context is time machines. It is at least as old as Rip Van Winkle, who went to sleep in the Catskills Mountains after drinking a dubious liquor and woke 20 years later, beard down to his chest but otherwise unaged. He missed the American Revolution against the British monarchy. His tale was told by Washington Irving in 1819. H.G. Wells used a similar theme in his 1910 book *The Sleeper Awakes* (an expansion of his 1888–1889 serial *When the Sleeper Wakes*). In 1897 Graham takes a medication to correct his insomnia and only wakes from an extended coma in 2100 to learn that during his long sleep his investments have multiplied by compound interest and his immense wealth now owns and commands most of the world, under the guardianship of deputies, the plutocratic White Council.

In the real world, that same general principle has given rise to a small band of optimistic futurists who have bought insurance to pay for the radical cooling and preservation of their bodies or at least brains after death. If biomedical research continues at its present accelerating pace, perhaps such cryonic suspension will hold their bodies in a state of minimal entropic damage, permitting specialists in coming decades to revive them and repair whatever injuries led to their death. Cryonics was publicized by science fiction magazines in the 1960s, but a version had already been used by Robert Heinlein in a novel serialized in 1956. Its charming title was prompted by his wife's observation that their cat declined to leave the wintery snow-bound house because he was looking for "a door into summer." It is ironic that such a door might exist precisely through the agency of intense cold below −130°C.

In the novel, innovative engineer Daniel Boone Davis takes this deep-freeze route from 1970 to the canonical sf future date, 2000. Later, he does it again, this time round with his cat Petronius the Arbiter. Neither exploit is genuine time travel; it just amounts to the technical miracle of blocking all metabolic activity—including, of course, consciousness—and then reviving the not-quite-dead person 30 years later. What is meant by the Time Travel Hypothesis requires an accelerated trip, probably through a higher dimensional hyper-space or perhaps a wormhole, implying that the world left behind sees no trace of the transition until it is over. What's more, it is usually implied that the transition can be made in the reverse direction.

Indeed this is what happens to Davis, who locates an eccentric university scientist in 2000 who is working on time travel, and has himself returned to 1970 after that first cryonic visit to the future—setting the scene for his second icy voyage forward to summer. Something vaguely similar occurs in Heinlein's 1964 *Farnham's Freehold*, where the eponymous individualist Alpha-Male (or is he?) and his family and a visitor are flung into the post-nuclear holocaust. Farnham and his new love are returned via time machine to nearly the same moment when a Soviet thermonuclear weapon is detonated almost on top of them. This time around the pair survive, and set up house and store for the few survivalists who also had the forethought to purchase a bomb shelter. (This kind of spacetime warpage via explosion, with no need for decades or millennia to pass in the encapsulated bubble of time, also accounts for the one-way voyage to *Utopia 239* by Stanley Bennett Hough—writing as "Rex Gordon"—published in 1959.)

Dan Davis cannot entirely avoid paradox in his re-entrant time loop. In the 1960s, he has designed, built and tested several household and workplace utilities, preparing to market them via Hired Girl, Inc., with his partner Miles Gentry and bookkeeper (and his fiancé) Belle Darkin. Flexible Frank, his household robot, sounds rather like an enhanced Roomba autonomous vacuum cleaner—in the real world, marketed in 2002. Betrayed by the two associates, who sell his company to a competitor, he is left only with a settlement and no rights to his inventions. Luckily, and perhaps disturbingly, Dan has captured the romantic attention of Ricky, the step-daughter of Belle, so he mails her his stock certificate and arranges to be cold-stored until the year 2000. Drugged by Belle, he is shanghaied to the future with a different cryonics company, and wakes to find himself virtually penniless and 30 years behind the times with his projects and engineering skills.

Now the apparent paradoxes kick in. Dan cannot locate Ricky, who by now could be in her early thirties; Miles is long dead; Belle is a shrewish drunken wreck. The company they sold his work to has collapsed, and his designs have apparently been implemented by Aladdin Auto-Engineering as "Eager Beaver"—a patent built on his prototype and signed by "D.B. Davis." Dan is no fool (except romantically, it seems), and understands that he must be going to travel back in time to undo the mischief and collect his cat Pete, abandoned by Miles and Belle. He tracks down Dr. Twitchell, an embittered genius rejected by his peers, who sends him back to several months prior to the betrayal.

Drawing on advanced knowledge acquired some 30 years in his now-future, he designs a drafting machine not altogether different from the computer-aided drafting and design devices now used everywhere in the real world (and in fact pioneered by Dr. Patrick Hanratty in the year after Heinlein's book was serialized). At last he absconds with the prototype Flexible Frank and his blueprints, and meets up with young Ricky at her Girl Scout summer camp (rather as Humbert Humbert did with Lolita in Nabokov's scandalous novel from 1955, but with only the most pure intent on Dan's part—although scarcely-pubescent Ricky agrees to go into cold sleep when she turns 21 and meet him in the future when they will marry). This happens. Fatal time loops are avoided—with the possible exception of Dan's advance knowledge of year 2000 engineering insights, but then some of those were realizations of his own 1970 insights.

Heinlein, author after all of "By His Bootstraps," is conscious of these potential problems. In the final pages of the novel, he has Dan reflecting on the oddities of time travel:

> Is the old "branching time streams" and "multiple universes" notion correct? Did I bounce into a different universe, different because I had monkeyed with the setup? Even though I found Ricky and Pete in it? Is there another universe somewhere (or some*when*) in which Pete yowled until he despaired…?

One paradox is unexplained, he notes, a line of print in the *Times* showing that he had arrived twice in the future, the second time earlier than the first. Could he have seen it and gone to confront himself? No, because he had no memory of doing so. But how could he have missed seeing it?

> …if I *had* seen it, I wouldn't have done the things I did afterward—"afterward" for me—which led up to it. Therefore it could never have happened that way. The control is a negative feedback type, with a built-in "fail-safe"… one of the excluded "not possibles" of the basic circuit design.

The universe of space and time as a circuit, with circuit breakers to prevent horrid paradoxes. That is very much Heinlein's view of reality, although it could go to the absolute extreme of rationalization to be found perfectly embodied in his famous short story "All You Zombies—" (1958), where a girl who does not know that she is a genetic mix-up with both uterus and penis and is made pregnant by herself as a male from her own future, giving birth to the girl baby who will grow in an orphanage up to be both of them. How could this possibly happen? Well, the confused man is inducted by an older version of himself, a bartender with a time machine who works for the Temporal Bureau precisely to prevent or set right paradoxes in time. This is a wonderfully neat solipsistic little masterpiece, but it leaves unsolved the origins of the Ouroboros loop which is the tangled world-line of the female/male (who wears a ring showing that mythic snake swallowing its own tail). In the bartender's future, does he create the Bureau that brings him/her into being?

Or is it just that Time Patrol-style agencies are on the lookout for potentially hazardous loops, to nip them in the bud then stitch them shut? Perhaps we are left with the fact that even so clever a scientific fantasist as Robert Heinlein could not resolve every last detail in any time looping, tail-biting tale.

1958/2002 *The Time Traders* Andre Norton; *Atlantis Endgame* Andre Norton and Sherwood Smith

If simplicity is a prime storytelling virtue, then Andre—originally Alice— Norton (1912–2005) clearly deserves her plaudits in fantastika: first woman Gandalf Grand Master of Fantasy (1977), first woman recipient of the Science Fiction & Fantasy Writers' Grand Master award (1984), and elected to the SF Hall of Fame (1997). Her fiction is cleanly written, plainly characterized, suitably for a novelist who gained her first and loyalist audience in children's and Young Adult forms of sf and fantasy. In her last years, most of her sequels were co-written with other women, and tends to the same "window pane" lucidity.

In a series of seven volumes that stretched over nearly half a century, with a three decade silence in the middle, Norton and much later her collaborators Pauline M. Griffin and Sherwood Smith dramatized the notion of time travel as a highly classified tool of the leading superpowers of the mid-century and beyond. They blend trade, exploration and exploitation of crashed starcraft millions of years ago, and covert war moves with the discovery of alien space-time travelers, the genocidal Baldies, who have malign plans for Earth. Interestingly, the production of these late collaborations were not without their own somewhat grim tussles. Reportedly, the penultimate volume *Echoes in Time* was meant to be written by Ms. Griffin, but the publisher insisted on Ms. Smith, who also wrote the final two books of another Norton series:

> When a friend of [Norton's] brought the title up in a conversation she directed them to remove this and the other three titles by Sherwood from her house. Upon her death she willed the copyrights to Sherwood because she hated all four books so much that she did not want her Estate associated with them. ...It is evident that Andre had little to do with them for they do not fit her style.[1]

---

[1] From "JW", at http://www.andre-norton-books.com/index.php/worlds-of-andre/series-by-andre-l-thru-z/time-traders-series/96-echoes-in-time . Sherwood Smith herself mentions some difficulties: "The road was not completely smooth… [Earlier volumes] were so very, very fifties. Cigar shaped rockets. Evil Russians. And the Baldies—the aliens with the big bald heads. These tropes, popular when I was a kid, had not aged well. But I had promised Andre that I would try to stay true to her original vision, and in

Operation Retrograde is a US top secret program (later to become the international Project Star). Studying and using time machines and ancient alien spacecraft, Retrograde seems somewhat like a mix of the 1940s' nuclear weapons Manhattan Project and the then not-yet-existing 1970s–1990s military Star Gate remote viewing program. To its hidden operational base, the youthful, angry and nonconformist petty criminal Ross Murdoch is dispatched by a judge who lets him choose this mysterious fate rather than standard criminal incarceration. Ross is trained as a Time Agent, and works with Apache Indian, Travis Fox, archaeologist Gordon Ashe, Polynesian Karara Trehern and her two telepathic dolphin associates, Tino-rau and Taua, and eventually the skilled, formidable young woman he marries, Eveleen Riorden.

In addition to transtemporal travel, the Agents and their support team also unlock the secret of time viewing—but this is a dangerous approach to information gathering because the Baldies with their advanced technologies and mental powers can use the signal to trace and destroy the user. Amid all the adventure huggermugger, Norton and Smith elaborate a model of time transport perhaps foundational of a now-familiar sf trope: the Time Gate. Such a device, given that name, lay earlier at the heart of Robert Heinlein's "By His Bootstraps" (1941). A television series helmed by Irwin Allen, *The Time Tunnel,* helped make this notion a catchword among young viewers in 1966–1967, nearly a decade after the first Norton Time Traders novel. A recent version was NBC's *Timeless* (2016–2018), with an historian, a soldier and a scientist teaming to roam through the past and keep it safe from interference from others.

Here is a brief description of Norton's temporal portal in the opening volume (updated in the 2000 Baen edition to replace "Reds" with "Russians" after the collapse of the Soviet Union, but after a further two decades still needing some tweaking due to the absence of, for random example, cell phones and married gays):

The transition itself was a fairly simple, though disturbing, process. One walked a short corridor and stood for an instant on a plate while the light centered there curled about in a solid core, shutting one off from floor and wall. Ross gasped for breath as the air was sucked out of his lungs. He experienced a moment of

---

our phone discussions, it became clear that, though she still read a great deal of anthropology and history for pleasure, she was not up on the rapidly changing world of high-tech. When I tried discussing ways of adapting the series for modern readers, it just made her confused and anxious" ("Andre Norton and Me," Tor.com, Feb 17, 2012, quoted here with her permission). Ms. Smith's changes and additions do seem suitable and helpful, but it is necessary to recall that Norton was then about 90 years old.

deathly sickness with the sensation of being lost in nothingness. Then he breathed again… (*The Time Traders*, 48)

The mechanism is not explained, but it does seem conclusively established that time transfer involves a considerable amount of gasping, sweating, nothingness and falling down. H.G. Wells's time traveler experienced an inaugural version of this *mal de temps*:

> I took the starting lever in one hand and the stopping one in the other, pressed the first, and almost immediately the second. I seemed to reel; I felt a nightmare sensation of falling… An eddying murmur filled my ears, and a strange, dumb confusedness descended on my mind.
>
> I am afraid I cannot convey the peculiar sensations of time travelling. They are excessively unpleasant. There is a feeling exactly like that one has upon a switchback—of a helpless headlong motion! I felt the same horrible anticipation, too, of an imminent smash. (*The Time Machine*, 19)

By the seventh and so far final volume, the process of shifting between eras is more advanced. On a trip back to the final days of Atlantis or Kalliste, known now as the Aegean island of Thera or Santorini which was wrecked by a monstrous volcanic eruption in the seventeenth century BCE, the Time Agents take their boat with them. The portal is established at sea by Russians in two ships carrying powerful energy sources, and seems to be a kind of mesh of spacetime fluctuations:

> …a sheath of blue light flickered over their hulls…. Deep below the range of human hearing, a vast bell toll… glowing mist surrounded them… power being deployed around them to wrench a 3600-year-deep hole in the universe…
>
> The water around their craft suddenly boiled without heat as the series of portal rods carefully spaced along both sides of the ship pulled power from the field now building between the Russian ships…. The mist began to flow inward toward a bright point of light. It was not a vortex, but straight lines converging on an infinity that flowed hungrily forward to engulf the boat, as though [Eveleen's] blind spot was expanding… There was now no sense of motion, only a sense of a physical violation so great it made nausea seem pleasure by comparison. (*Atlantis Endgame*, 46–7)

This kind of unpleasantness, like a rite of passage, used also to be an intrinsic part of the routine when traveling through hyperspace in your starship. Bearing all this in mind, you might be inclined to stay at home.

For Norton, like the Area 51 conspiracy theorists of today's talk-back radio, the secrets of space and time travel did not emerge *ab initio* from human research labs but rather by reverse engineering alien craft. In the Time Trader series, these mostly crashed thousands of millennia ago in locales scattered about the globe but were then buried in glacial ice or the accumulated rock of ages. As Ross Murdoch is told in his first encounter with these phenomena,

> So the crackpots were right, after all… They only had their times mixed… Flying saucers… It was a wild possibility, but it was on the books from the start… This was insinuated once, by a gentleman named Charts Fort, who took a lot of pleasure in pricking what he considered to be vastly over-inflated scientific pomposity. (*The Time Traders*, 139–40)

As we shall see in our final section, this might not be as crazy a notion as it seems even today in a world where the scientifically endorsed Search for Extraterrestrial Intelligence (SETI) listens to the stars for pulses of ancient radio or optical messages, and military aircraft and radar continue to report the presence in the skies of impossibly agile craft not of any known source.[2] UFOs as time machines? That would be an unexpected corollary of the Time Machine Hypothesis, even though Andre Norton was postulating it, in fictional form, more than 60 years ago.

1958 *The Big Time* Fritz Leiber, and 1983 *Changewar*

As noted above, Jack Williamson got there first with a novel about a conflict for realization between two competing parallel worlds, but his landmark narrative is damaged by its clumsy voice. The dramatics of Fritz Leiber's 1958 Hugo-winner and its sequel "No Great Magic" (included in the 1983 collection) are very far from clumsily wrought. "Dramatics" must be taken literally here, as well as symbolically, because both fictions take place in a sort of space-time vacuole that is the home and stage of a troupe of actors.

Leiber knew this fraught arena well; his father, Fritz senior (1882–1949), was a Shakespearian and film actor (starring as Caesar in the 1917 silent movie *Cleopatra*, with Theda Bara), and in his youth tall, striking Fritz Jr. (1910–1992)

---

[2] See, for example, the story that made headlines in December 2017 when it was revealed after years of denial that the Pentagon had continued studying UFOs: https://www.washingtonpost.com/news/retropolis/wp/2017/12/18/the-government-admits-it-studies-ufos-so-about-those-area-51-conspiracy-theories/

was also an actor. His writing echoes many of the great classics of the stage, whether set in mythic sword-and-sorcery land of Nehwon (nowhen in reverse), or the hidden endless conflict of the Changewar. His opening of *The Big Time* (a term borrowed from theater, music, and sport, but here warped into the largest canvas conceivable) is direct:

> My name is Greta Forzane. Twenty-nine and a party girl would describe me. I was born in Chicago, of Scandinavian parents, but now I operate chiefly outside space and time—not in Heaven or Hell, if there are such places, but not in the cosmos or universe you know either…. I have a rough-and-ready charm… I need it, for my job is to nurse back to health and kid back to sanity Soldiers badly roughed up in the biggest war going. This war is the Change War, a war of time travelers—in fact, our private name for being in this war is being on the Big Time. Our Soldiers fight by going back to change the past, or even ahead to change the future, in ways to help our side win the final victory a billion or more years from now. A long killing business, believe me. (*BT* 151)

These combatants and their support teams represent two secret forces whose arena is scattered across time and space, the self-named Spiders and Snakes, yet there seems no clear ideological or commercial motive for the end-less war. "Our side" in Greta's case are the Spiders, and her small corner of the Big Time contains an old ham, Sidney Lessingham, who knew Shakespeare and speaks a kind of Elizabethan patois, a World War I British soldier, a steamboat cardsharp with exquisite manners, a Nazi who seems to be Greta's main lover, a rather octopus-like alien, a drunken physician, and others one would not expect to be aligned. Is Greta a whore, or just a good-time gal? In the sequel, which follows certain catastrophes not developed in other stories, Greta has a serious head injury that has left her "A&A," Agoraphobe and Amnesiac. Her confused misunderstanding of much that occurs eases the reader into a realm where—for no explicated reason—the troupe perform *Macbeth* for an audience that includes Queen Elizabeth I and her courtiers.

Several lesser tales relate to the war between Snakes and Spiders, or simply to the mysteries of time travel. Perhaps the most well-known is "Try and Change the Past" (1958), which tracks a womanizing creep living on his embittered wife's money through his recovery after death by murder to the condition of Doubleganger. This is a somewhat ghostly Resurrected existence sustained until he agrees to join the Change War, but he is determined to use his new time-flexible condition to change his own past. Taking the pistol his wife will use to shoot him between the eyes, he empties the rounds—but dies anyway, because he has removed only the ghostly bullets available to his condition. He tries again, but finally, having killed her in turn, he is slain by a meteor of precisely the right size and density to stand in for the 32-caliber bullet that destroyed his brain. The lesson: "Change the past and you start a wave of changes moving futurewards, but it damps out mighty fast.... So how's a person going to outmaneuver a universe that finds it easier to drill a man though the head that way rather than postpone the date of his death?" (*Changewar*, pp. 1, 11).

But if this shaggy-god jape tells us that time's course cannot be changed, why are the two great powers outside space and time doing their best to gain victory? And what would constitute victory? In Asimov's Eternity, it is the constant effort to trace the source of a chain of causality leading to an undesirable outcome—starflight, especially—and nudge this *jonbar* point in a preferred direction. That works until Andrew Harlan and his far future guide obliterate the whole hubristic project before it even begins. Leiber's version is more like an endless joust between two Eternities, equally matched, and without any understandable stakes.

Sid's Place of Entertainment and Recuperation is "midway in size and atmosphere between a large nightclub where the Entertainers sleep in and a small Zeppelin hangar decorated for a party" (*BT* 152). It hangs in the Void and is assailed by the Change Winds. It is easy to detect in this brew elements of Leiber's other literary interests: horror, and the fantastic. Yet access to this extra-dimensional *cul de sac* is achieved and sustained by two highly advanced items of technology, the Major Maintainer, a "great dull-gleaming jewel, covered with dials and green-glowing windows" (*Changewar*, 197), and the Minor Maintainer, which can do such tricks as inverting objects through a fourth spatial direction, so that a left hand glove will fit a right hand. When a nuclear weapon is introduced into the Place by one of the cast, and no way to dispose of it or disable its settings, the plot tightens.

In short, there is no explanation of time travel provided, even in handwaving form—just a well wrought character study of people from different eras interacting under threat and pressure in a frighteningly pervasive and murderous standoff shaping our reality. Death is no longer a guarantor of extinction and surcease. Shakespeare himself is shaken by his viewing of *Macbeth*, and stands in for us all:

> What may this mean? Can such things be? Are all the seeds of time... wetted by some hell-trickle... sprouted at once in their granary? Speak... speak! You played me a play... that I am writing in my secretest heart. Have you disjointed the frame of things... to steal my unborn thoughts? Fair is foul indeed. Is all the world a stage? Speak, I say! Are you not my friend Sidney James Lessingham of

King's Lynn... singed by time's fiery wand... sifted over with the ashes of thirty years? Speak, are you not he? Oh, there are more things in heaven and earth... aye, and perchance hell too... Speak, I charge you! (*Changewar* 195–96)

There is no reply.

1958/1992 *The Ugly Little Boy* Isaac Asimov and Robert Silverberg

If Dr. Asimov was one of the first science fiction writers to explore the massive sociological impact of time travel in *The End of Eternity*, he was also notable in telling a close-focus humanist tale, "The Ugly Little Boy," about a Neanderthal boy and the near-future woman he came to call Mother. Asimov was then 38 years old, famous for his galaxy-spanning *Foundation* trilogy but not at all known for stories aimed at tugging the heartstrings. (One of them was emotionally fraught, admittedly: the 1956 time viewer story "The Dead Past.")

Nearly the same number of years later, not long before his death, he allowed his friend Robert Silverberg (b. 1935) to expand and deepen the story into a novel published in the US under the original title and in the UK as *Child of Time*, borrowed from its opening poetic epigraph. Here we acknowledge the novella as the seed of the collaborative work, which incorporates the Asimov text but enriches it with Silverberg's additional twenty-first century characters and motives plus an entirely new thread revealing the People, 40,000 years prior, whose child was snatched deliberately into the future as if he were a fish on a line or a dog caught in a trap.

The People live in a time where their sub-species, *Homo sapiens neanderthalensis*, is in fear of their new rivals, the *Homo sapiens sapiens* who encroach on their holy sites and traditional hunting realms. Silver Cloud is an ancient shaman, and his aging rival She Who Knows was formerly "the beautiful slender Falling River" (p. 2). Minor figures include Broken Mountain, Tree of Wolves, and so on. These are not brutish sub-humans, but we know—and as they fear—they are doomed in this contest for survival by their mismatch with a changing environment.

In approximately our time, Dr. Gerald Hoskins is Chief Executive Officer of an expensive project to use a newly built Stasis generator to reach into the past and draw out sample rocks, primitive artifact, and the child, Skyfire Face (as the People name him for his "jagged lightning-bolt birthmark" (71). (Curiously, this was published half a decade in advance of Harry Potter's similar and iconic scar.) Hoskins has interviewed three drastically different candidates to act as the child's nurse and custodian within the Stasis field, and the job goes to Miss Edith Fellowes rather than Marianne Levien—a "real tiger" later associated with do-gooder critic Bruce Mannheim, a kind of Peter Singer. Initially repulsed by the Neanderthal's terror and apparently deformed body and face, Fellowes is soon won over and names him, touchingly if somewhat absurdly, Timmy.

It is quickly established that the Stasis bubble is safe to enter, with only a momentary "queer sensation." The factor stressed with especial force is that nothing—including Timmy—can be removed safely (or, more explicitly, cost-effectively) from the bubble

> Nothing from the past can be removed from its Stasis bubble and there's no way around that. It's a matter of maintaining the balance of temporal potential… An energy-conservation problem. What comes across time is traveling across lines of temporal force. It builds up potential as it moves. We've got that neutralized inside Stasis and we need to keep it that way. (129)

Teams of researchers from diverse disciplines cluster around the bubble to observe the child, now frantically bonded to Miss Fellowes. This is momentous and enthralling for scholars, but the citizens at large are quickly bored. Hoskins, whose doctorate was on the nature of time via mesonic detection, hopes to find some way to detect and grapple with objects nearer than the present 10,000 year limit—preferably a recognizable person who speaks a language linguists can cope with. Only Fellowes has had the intimacy, kindness and persistence to teach Timmy the rudiments of English. It is clear that he is a bright child, although as the months pass his limitations start to manifest themselves.

Can history be changed, using Stasis to reach into the past? Possibly, but changes damp out in a converging series. "The amount of change," Hoskins notes, "tends to diminish with time and eventually things return to the track they would have followed all along."

"You mean, reality heals itself?" (137).

This is the premise common to much sf speculation about time travel. In this case, self-healing will be aided by the decision to send the now doubly-deracinated child home in the far past, even though it is likely that his family will have moved away by now—and he lacks the basic education of a Neanderthal. Miss Fellowes is aghast, and makes plans to take him into our world, freed from the Stasis bubble even though taking him out will see "a year's supply of energy gone in half a second." Hoskins stresses this: "an object returning to its place in the space-time matrix generates such powerful forces in its immediate vicinity that it takes with it anything that's nearby. The mass limitations seem to apply only in the forward direction." (134). Timmy "was taken from the far past. It moved across the timelines and gained temporal potential. To move it into the universe—*our* universe, and into our time— would absorb enough energy to burn out every line in the place and probably to knock out power in the entire city" (155). Since the work is conducted by a commercial enterprise, Stasis Technologies, this is really the deciding factor.

Meanwhile, Hoskins is excited that the techniques are now improved to the point where the mass limit for cargo from the past is no longer a mere 40 kg but closing in on 100 kg, with a 10,000 years envelope and better predicted. To make optimal use of the finances and existing work spaces (264), Timmy must either be euthanized like any other test animal (one monstrous suggestion) or sent home and abandoned there to his fate.

Project Middle Ages is a success, nabbing a man from historic times. Now their process will carry new knowledge of antiquity into the twenty-first century, "an incredible intellectual treasure." Now three years older, Timmy is prepared for his return. Miss Fellowes fails in her attempt to take him out of the Stasis bubble, and instead enters herself and trips the activating circuit (283). The finale, like many previous interlineated chapters, follows the *Homo sapiens neanderthalensis* clan to which Timmy is returning. They are threatened by *Homo sapiens sapiens*, "The Other Ones," but both groups watch in awe as the Goddess the clan worships appears with Timmy. (285–290). The shamans of the Neanderthal People, male and female, kneel, as do the approaching Others]:

> How strange her face was! And—though it was an Other One face… how beautiful, how tranquil! She was smiling and her eyes were shining with joy! The Goddess stood waiting, holding the boy Skyfire Face still in her arms… A golden light seemed to stream from them both… Everybody side-by-side, all thoughts of warfare forgotten, one by one kneeling in the snow, looking up with wondering eyes to pay homage to the shining figure with a smiling child in her arms who stood in the midst like a harbinger of springtime and peace.

If a movie is ever made of this book, there won't be a dry eye in the house at that climax. It must be admitted, though, that a further three decades or so after Silverberg expanded Asimov's intentionally sentimental plot, viewers might be rather more skeptical that a group of People threatened by Cro-Magnons would find Fellowes' "Other One face" anything but ugly, and her body unpleasantly configured. That is surely one of the major risks of time travel.

7

Behold the Time Machine

1962–1969 *Times Without Number* John Brunner

In 1988, Spain is celebrating the four centuries of its victory since her invincible Armada crushed England's fleet, making the monarchy the world's dominant empire. In this alternative history by John Brunner (1934–1994), a major player in Spain's triumph was and remains the Society of Time, of which time traveler Don Miguel Navarro is a Licentiate in Ordinary and loyal subject of His Most Catholic Majesty Philip IX, *Rey y Imperador*. For a short novel of peril compiled in expanded form in 1969 from three linked stories published in the British magazine *Science Fiction Adventures*, it remains a thoughtful if neglected study of time paradoxes and opportunities.

In this allohistory, even late twentieth century *faux*-egalitarians in Jorque (York in our world) own black slaves—"Guinea-men" and "Guinea-girls"—and a Mohawk prince rules New Castile, while the Inquisition decides on breaches of the law by those who abuse time travel regulations. Commonplace arguments are rehearsed: is the record of events as steady in the face of a retrocausal event as the flow of a stream across millions of stones despite the shifting of a single stone, or might that stone's dislodgement be a tipping point that is magnified into an avalanching flood that diverts the river's course? An example suitable for the four-hundredth anniversary season is voiced: might the Armada have lost had a fierce storm failed to soak the British fireships, thus permitting Spain's galleons to burn? A time traveler could not change the course of a major gusting storm, but might be able to persuade military and naval leaders to alter the time table of attack sufficiently to avoid the weather change.

© Springer Nature Switzerland AG 2019
D. Broderick, *The Time Machine Hypothesis*, Science and Fiction,
https://doi.org/10.1007/978-3-030-16178-1_7

For the purposes of this study what is most interesting is Brunner's development of a time travel theory and perspective, set within a culture divergent from our own by crucial centuries. Science in general is thwarted, so that the notion of someday flying to the Moon is shorthand for the simply ridiculous. Flat photographs exist, but apparently no motion pictures. No gasoline vehicles, just horseback, wagon or coach. Yet somehow that world has had time travel since its discovery nearly a century earlier by Don Carlos Borromeo. This experimenter was so disturbed by the likely consequences if the technology became widely used (it was built readily from silver and iron) that he handed over its disposition in the Empire to his newly founded Society of Time and in the Confederacy of the East to the Temporal College. (Spain itself had been ceded to Islam.)

A Jesuit, Father Ramón, heads the Society from Londros and discourses briefly with Don Miguel about the varieties of time. Navarro had spent 3 years of higher education as a Probationer studying temporal physics, wrestling with the relationship between familiar *substantive* time, in which one measured out one's daily life, and hard-to-grasp *durative* time in which one experienced events during a time-journey, and he had written his graduation thesis on the subject of so-called *hypertime*, the barrier which prevented a time-traveler returning from the past from going any further futurewards that the moment "then" reached by the apparatus which had launched him.

> But all these were as nothing compared to the hypothetical complexities of *speculative* time, in which events would be otherwise than as history recounted. (32, emphases added)

These distinctions gesture toward a more mature and complex understanding of time than Wells's Traveler possessed, with his single extra dimension, or Asimov's doubled temporal dimension. It is closer, in science fiction, to James Blish's coolly intellectual quasi-Jesuitical operations with spacetime in the composite *Cities in Flight* (1970), multiple superposed strands of reality in *Jack of Eagles* (1952) or indeed demonology in his linked novels *Black Easter* and *The Day After Judgment* (1968, 1971); none of the latter, though, deals with time travel. Brunner's classifications are enacted in the novel, not merely listed. It is rather striking, and amusing, that the fourth brand of time forms the basis of what is felt to be a scarifying insight by pious Catholic Navarro: initially, by guessing that in a world of unfaithful time travelers unconstrained by the Society's rules, might agents of evil "plot journeys back into time, with the intention of undoing the good consequences of the acts of others?" Even "to deliberately corrupt the great men of the past?"

Father Ramón is relentless. The historic influence of evil "might be the working out of just such interference as you suggest. Some theorists have even argued that the fall of the angels hurled from heaven may have been a plunge through time rather than through space" (48). Atheist Brunner, like atheist Blish, enjoyed lobbing such grenades on to the playing fields of theology. But the Jesuit is not finished. What makes an act of free will *free*? Why, "that all possible outcomes be fulfilled." For Navarro, this is an ontological horror story. Ramón hammers home the final logic, which amounts to the Many Worlds theory advocated in our history by Hugh Everett and in their world seemingly validated by visits to many alternative histories endorsed or expunged:

> If there is free will—and we hold *a priori* that there is—all our opportunities for decision must conclude in just so many ways as there are alternatives. Thus to kill and not to kill and merely to wound more or less severely—*all* of these must follow upon a choice between them. (49)

I am fairly surely this is sophistry, since the quantum multiverse is shaped by *probability measure*, or share of possible outcomes.[1] One might expect *some* proportion of heinous actions without being driven to the nauseating conclusion that they are as numerous as neutral or benign Everett universes. Don Miguel, ignorant of such analyses, feels the ontic certainties beneath his boots swirling into nothingness. "…if this is true, it scarcely seems to matter whether we interfere or not! We ourselves may be only a fluid cohesion of possibilities, subject to change at the whim of someone who chooses not to obey the rule of noninterference." The Jesuit solemnly agrees that this is so, the price of free will.

All of this is contained in Part One of Brunner's thoughtful novel of a Spanish imperium at least as ponderous and pompous as our own, replete like ours with sly and self-serving creatures with power or desperate for it. In Part Two, the steam and seriousness rather escape, with a murderous invasion of Amazons from a rogue history. Now we seem firmly relocated to a landscape of universes where all possible outcomes must be fulfilled as Alternative Histories, expanding the original premise that a principal goal of the Society of Time is to observe, never to change history. Luckily, by creating a brief closed timelike loop going back in time to block the attack, the regime is

---

[1] Alexander Vilenkin "A quantum measure of the multiverse" (2013): https://arxiv.org/pdf/1312.0682.pdf

spared, the King is no longer dead—and evidence is thus strengthened that the Jesuit's free will theorem is correct.

By Part Three, the final section, the long-established Treaty of Prague governing criminal use of time travel seems under threat by the Confederacy of the East (between Europe and Cathay, or China) who are/were mining precious ores a millennium earlier in California. Don Miguel and the Jesuit theoretician are sent back to deal with this most terrible of potential tipping points. Even after this serious risk is sorted out, the larger concern haunts Navarro and the assembled General Officers. It seems likely that the planner of this crime intended it as a distraction from his true goal: to change the probability of victory by the Spanish Armada. Thereby inciting wars among the many weaker nations of the world.

To prevent this immense genocide of all living humans and their replacement by the denizens of an entirely different future, Navarro is translocated to Cadiz in an attempt to track his foe before something is done in 1588 to sabotage the galleons of the Armada. Everything goes awry. I leave these intriguing details to the reader. But the novel's last bitter irony (avert your gaze if you wish to avoid this final spoiler) is Navarro's stranding in a speculative time track that, with his history's obliteration, isolates him amid the bare-limbed young women and men and towering buildings and stiff winged things high in the heavens. Marooned in Nueva Jorque, our New York, a world that plainly does not yet support the existence of time machines, he swears it will never learn from him how to make one. In Brunner's final elegiac words, "now the most isolated of all the outcasts the human race has ever known, [Don Miguel] walked forward, into the real world" (156). Or at any rate, into our world so infinitely detached from his own lost spacetime that it can be regarded as the only real world.

1966–1969 *Behold the Man* Michael Moorcock

One of the commonest reasons offered for the impracticality of time machines is the absence of conspicuous time travelers, especially at historically memorable events. Perhaps the most plausible and certainly convenient explanation is that no time traveler can go back in time further than the first moment the machine was activated. If that is true, then perhaps visitors from future times near and far will start to pile up in, say, 2050, or 3124, or closer to the year million. But let us suppose that time machines will have the power of creating wormholes (say) that let daring researchers emerge into the past. Where would they choose as their favorite destinations, and what would happen if they got there? If it proves impossible to *change* the past, then at least travelers might already have *contributed to* known history.

In nations with a Christian cultural heritage, it is often asked, perhaps blasphemously, where the onlookers were who will surely have gathered at the crucifixion of Jesus, or perhaps worked some medical miracles to bring his corpse back to life. In a notable short story, "Let's Go to Golgotha" by Garry Kilworth (1975), most of the onlookers at the trial of Jesus, recorded in Scripture as having chosen to spare the criminal Barabbas over Christ, were suitably disguised time voyeurs. The first voice calling for the death of the messiah was from one of them, quickly joined by the braying and wildly excited mob of time travelers who had been advised in advance to do so lest they change history. Instead, they created that history; there were no Jews present at all, only nosy travelers.

There is another, far more unsettling possibility that's been explored as an sf trope: what if a traveler *replaced* a historic figure. This happens in Poul Anderson's very good early Time Patrol story "Brave to Be a King" (1959), in which patrol officer Keith Deninson is obliged to adopt the role of Cyrus II the Great King of Persia, abandoning his twentieth century wife. This necessary imposture is corrected after he has spent many years ruling Persia, married a woman he loves dearly, fathered her children. Read sympathetically, it is a heartbreaking tale, and an intriguing exercise in apparent time paradox. But another story in the same vein, Michael Moorcock's novella and later novel "Behold the Man," is far more audacious and won him the 1967 Nebula award. Now the traveler is a neurotic Jewish incompetent, Karl Glogauer, whose sharp-tongued lover Monica insists that the Christian mythos was cobbled together from earlier legends and political contrivances. A brilliant but disreputable scientific acquaintance testing a time machine of his own devising sends him back to the year 28 to observe the historicity or otherwise of Jesus.

In this quite brilliant and well-written story, surely judged by the Christian faithful as one of the most wicked ever published, the real Jesus turns out to be an illegitimate "congenital imbecile" (101), with "a pronounced hunched back and a cast in its left eye. The face was vacant and foolish. There was a little spittle on the lips... It giggled as its name was repeated" (98). His purported father is a feeble fellow tricked into marriage by not-so-virginal Mary, who was by now "tall and beginning to get fat. Her long black hair was unbound and greasy, falling over large, lustrous eyes that still had the heat of sensuality" (97). Presumably the Papal rite mentioned in Brunner's *Times Without Number,* whereby every newly-elected pontiff is taken to observe Jesus, happens in a quite different universe. In the perhaps inevitable regressive

Oedipal outrage (for Karl will now, of course, be trapped into adopting the role of the Savior Jesus),

> The lust in her big body was becoming hard to control. She hitched her skirt up to above her calves and spread her legs apart when she sat on the stool near him… Then she was on him, her hands tearing off his rag of a loincloth, her fingers on his genitals… He gasped and drew up her skirt, driving his fingers into her… (102–103)

to their consummation, as "The idiot stood in the doorway looking at them, spittle handing from his chin, a vacant grin on his face" (103). Half a century after this story was published it retains its power to shock, but not in any blustering or sniggering way. When Karl subsequently roams Judaea, more than half mad, at first just one of the numerous deranged prophets of the land but growing into his fated role, he "told himself that he was not changing history; he was merely giving history more substance" (118).

He attracts followers, bestows psychosomatic cures, and finally is seized by the Roman procurator Pilate and the Jewish Tetrarch Herod Antipas, each for his own motives. The crucifixion follows, and in a clever moment (one might wonder if this was the very source of Moorcock's idea for the story) he cries out as he hangs near death the words recorded in the New Testament as "Eloi, Eloi, lama sabachthani" ("which is, being interpreted, My God, my God, why hast thou forsaken me"—Mark 15:34) but is in fact the even more dismaying realization "It's a lie—it's a lie" (142–143). His corpse is soon stolen by doctors eager to dissect it in search of the mysterious properties that had given him healing powers.

Throughout, there are premonitory signals of Glogauer's suffering path toward death, from his emotionally absent parents to his repeated brutalization and sexual confusions as a child and adolescent finally acted out as *faux*-suicidal performances that convince neither his mother nor irritated Monica. A developing obsession with Jung's ideas draws him toward archetypes, so the doomed god-man is a role he must find irresistible. All of this can be accounted for in a time travel narrative with no unsupportable difficulties, until he starts to enact his memories of the Christian documents, choosing twelve disciples because that is what the Bible reported, making certain one of them was named Judas and ensuring his "betrayal" to the Romans by sending that luckless fellow to lay false charges of treason against him.

These are paradoxical time loops in terms of their rootless causation, like Bob Wilson's handy notebook (in Heinlein's "By His Bootstraps") filled with translation tips written later after he had learned to speak this future language with its help. Or one might, in Karl Glogauer's unique case, choose simply to regard this re-entrant causal loop as a divine miracle…

1969 *Up the Line* Robert Silverberg

Screwing the allegedly pure Mother of the Son of God (in Moorcock's *Behold the Man*) or oneself (as in Heinlein's "All You Zombies" or David Gerrold's *The Man Who F…ed Himself*—where the title's verb is "folded" but we know what is meant) are leading candidates for a science fictional Épater les Bourgeoisie Time Travel Sex Award. The same can't really be claimed for *Up the Line*, which is a merry escapade: a young man's very distantly incestuous adventures with his great-great-multi-great grandmother, the passionate Pulcheria.

About a decade before writing this novel, Silverberg (b. 1935, now a Grand Master for his stylistically eclectic and ambitious later books) turned out with incredible speed some hundreds of porn paperbacks under any name but his own. He picked up bad habits from this factory-line productivity, but soon abandoned that lazy if lucrative process for the harder taskmaster of literary sf. Some of these "new Silverberg" novels from the mix-1950s and later dealt with time travel—*The Time Hoppers* (1967), *Hawksbill Station* (1968), *The Masks of Time* (1968), and the retrocausal *The Stochastic Man* (1975), but perhaps the best liked is this time-looping romp through Byzantium with Time Courier Judson Elliott II in the early sixth century.

We meet Jud (b. 2035) with his incomplete history doctorate from Harvard, Yale and Princeton, in 2059 aged 24 as he decides in his jaded weariness to join the Time Service, although its more risky arm, the Time Patrol, does not appeal. Introduced to this role by a very black street-wise Courier named Sam, who is a decade his elder, Jud is swiftly taken on a series of trial trips wearing his new truss-like device (which will allow him to move back and forth in time as long as he keeps the phlogiston charged up). Accepted, he is lectured at length along with his group of eight newbies in the headache-inducing oddities and paradoxes banned and avoided or corrected, taking all this to heart. What could possibly go wrong?

A great deal, as it turns out. So much, indeed, that one must wonder how time travel could ever work without inducing catastrophes at every turn. Explanations are offered, but when those are questioned the instructors always get flustered and beg off. The suspicion grows that it is Mr. Silverberg who is avoiding the more perplexing questions, but that is acceptable in a rambunctious, sexy science fiction tale because, when all is said and done, none of these logic problems have yet been solved by the finest philosophers and mathematicians. Consider, for example, the Cumulative Paradox classically invoked at Golgotha. Every scene of sufficient drama, especially the death of a charismatic leader, should draw tourists from many centuries or millennia yet to come.

Jud reflects on the absence of thousands of crowding time tourists at the assassination of US Senator Huey Long in Baton Rouge on September 10, 1935. Well positioned by their trainer, who is masked enough not to set off rumors of identical triplets, his small team notice other clumps of travelers carefully maintaining their distance and even their gaze. But why is there not a huge horde? Could there be a Law of Conservation of History? asks one of Jud's team, rather as Stephen Hawking proposed many years after Silverberg's novel was published. "We do not care to risk it… We employ the Time Patrol to make certain that everything will happen in the past exactly as it did happen, no matter how unfortunate…" Why? Because the consequences could be far more horrible. Still, there remains the immense difficulty that if as few as a hundred sight-seers from any given year visit the Crucifixion of Jesus, there will be an extra ten thousand converging there by the middle of the twenty-second century and by the early thirtieth an impossible swarm of one hundred thousand would gather there, and thereafter millions (28–30). This would certainly be a matter of historical record, which so far it was not. Could the records change as time passed? Perhaps, but if so it would already be the case, since the convergence focused on an event that had already happened.

Up the Line has the virtue of being a funny book, where the amusement largely arises from black humor. Jud is drawn into the leisure activities of a wealthy Courier whose detestation of his cold, cruel father, grandfather, and earlier males of his lineage leads him to seduce as many of his female ancestors as he can trace and visit in the past. His mother as well? asks horrified Jud. "I draw the line at abominations," he is told (102). Jud falls obsessively in love with his gorgeous 17-year-old ancestor Pulcheria, and she with him. You just know this can't end well.

Numerous time tangles and potential doublegangings await imprudent, love-blinded travelers, but worse is in store. There's the Ultimate Paradox, in which a change might be engineered so drastic that the Benchley Effect that undergirds time travel is never discovered, thus abolishing the Time Patrol so that "time-travel becomes its own negation" (38). But wait, what of Time Service operatives in transit when this terrifying event occurs? (That is something already explored by Poul Anderson in his *Guardians of Time* sequence.) This is dubbed the Paradox of Transit Displacement: "a time-traveler in transit is a drifting bubble of now-time ripped loose from the matrix of the continuum, immune to the transformations of paradox" (38).

Oh? But why should they be immune? Because this would produce the Ultimate Paradox, even more paradoxical than the Paradox of Transit Displacement: "by the Law of Lesser Paradoxes, the Paradox of Transit Displacement, being less improbable, holds precedence." Such a fate actually befalls one character near the end of the novel, who remains immune to extinction until the moment he returns to his future now-base. So too does Judson Elliott III. "You're a residual phenomenon," he is told, "a paradox product, nothing more" (247). When he vanishes, everything he had previously done to mess things up is reverted by the Time Patrol. Poor Jud is not only dead, but his entire history (along, paradoxically, with the unfinished memoir we have just finished reading) is obliterated as well, has never been.

Silverberg is playing with us, of course, ironizing the generic handwaving typically employed to resolve time travel incongruities. Still, stranger claims have been advanced by quantum theorists and theologians, so something of this sort might turn out to be surprisingly accurate—if time machines are ever developed in portable, wearable form.

1970 *The Year of the Quiet Sun* Wilson Tucker

When he was 17, Arthur Wilson "Bob" Tucker (1914–2006) published his first fanzine—one of the amateur magazines that helped knit together the geographically dispersed readers of science fiction until the explosive arrival of cheap computers and the Internet, when blogs replaced almost all the by-then-classic self-produced paper'zines mailed locally at first then often internationally. Bob Tucker, one of the first and soon majorly influential sf fans, went on to write crime and sf novels while working as a film projectionist. It is intriguing that this pioneering fanzine was titled *The Time Traveller*, and that nod to a sub-genre launched by H.G. Wells less than two decades before Tucker's birth recurred as a theme in the notable novel under discussion—nominated for both the Hugo and Nebula awards—and several other sf works such as *The Lincoln Hunters* (1958).

These time travelers do not build their craft at home or in a commercial or academic laboratory. The first TDV, or Time Displacement Vehicle, is constructed and tested under the government auspices of the US Bureau of Standards in a heavily protected concrete structure, and powered by a dedicated nuclear reactor outputting 500,000 kW hours. The equipment is rated for 20 years of continuous service, and all equipment will be replaced every 25 years or sooner. Because the mass of the TDV and its single passenger makes heavy demands on the system, travelers are obliged to go into the future wearing only briefs. Suitable garments, food and water, camera, recorders and other provisions will be set out for their use on arrival at a prespecified future date, and they are permitted only 50 h outside the bunker before returning.

Many sf stories assume that experienced hours spent in the future will be matched by an equal time lost on return, but in this novel the traveler will be back in his home time (all three are male in this 1978 test sequence, going to 1980) after exactly 61 seconds. The initial test killed nine technicians and the traveler, because the homing TDV tried to materialize on the same spot and at the same moment it had left, with predictable explosive consequences. (One might think this sort of error could not possibly happen, but then one recalls the $125 million climate orbiter that smashed into Mars in 1999 because the engineering staff had confused Imperial and metric units.) Since that tragic and humiliating error, the system has been fully recalibrated, and monkeys are sent to the future and back without harm. After shorter leaps into the future, the plan is to explore the shape of things to come in 2000.

The temporal investigators are inducted by a lovely and competent Bureau specialist, Kathryn or Katrina van Hise, and comprise Air Force Major William Moresby, a seasoned warrior and evangelical zealot in his mid forties, a Navy Commander of thirty, Arthur Saltus, who is instantly smitten by Katrina, and the demographer and futurist Brian Chaney who is no less taken by the young woman's fashionable "delta pants" and transparent bra while too insecure to make advances. He is with the other two because of his futurological projection of the next few decades in the US but also for his recently published translation of ancient scrolls, which had caught the eye of Gilbert Seabrooke, the Director of Operations. His book had unleashed vituperation from many who share Moresby's beliefs. All of them, rather oddly, including the expert Chaney, keep mentioning the Book of Revelations; actually, that Biblical text has no terminal *s*, because it is named for the revelation it announces and not for the many hallucinatory images and apocalyptic claims it provides to the credulous and readily infuriated.

The Major, who is one of the latter, is convinced that the year 2000 will be the end of days. Chaney tries to explain that these purported divine visions are actually *midrash* or fables and parables, recognized as such by those who wrote and heard them. By the novel's disturbing end he finds himself wondering if maybe Morseby might be right.

As it turns out, both Brian's forecasts and Tucker's are mostly far from the mark, fortunately. The core driver is race war, trigger by the continued cruelty of the white communities toward increasingly ghettoized and walled-off blacks.

In the real world, this does continue, but not to the degree Tucker supposed likely (or rather, that his novel needed to project as likely). The time explorers range in the immediate proximity to their machine, the two warriors boldly, while Brian mostly lurks in the local library, copying recent military and economic data. Their first target was altered to satisfy the weak President Meek's demand to know if he would survive the next election (he will, after ratcheting up the hatred and incipient, paranoid race war afflicting the USA).

Here are two absolutely crucial spoilers, so look away if you have not yet read this quite important novel. From the beginning, Chaney has found that the TDV is not performing according to the self-satisfied predictions of its designers. Time of arrival in the future is out by minutes and then longer, as confirmed by the other military men. Their ventures into turbulent conflict are portrayed with sharp conviction. Although Tucker never served in the military during the Second World War, some of his scenes suggest fighting experience but then he spent much of the night watching the movies he projected (few of which, admittedly, would have been highly accurate).

In Chaney's final advance into the new millennium, the gradual deterioration of the bunker has reached serious proportions. Dust is everywhere. Scavengers have been breaking in and stealing food and clothing; the power is completely dead. His final discovery is that the reactor has failed, shutting down the link needed to get him back to the twentieth century. The base is completely surrounded by wrecked, burned vehicles and barbed wire. The staff have withdrawn with no messages left for him. Throughout this series of unfortunate events people in the increasingly disturbed streets have watched him with trepidation; now he sights a young family on the other side of the fence and they run in terror.

Finally he find the aged Kathryn, and her husband Arthur's grave, and meets her grown children. The daughter, of course, is the image of her beautiful mother, so Brian is smitten all over again, making this a happy ending of a rather bleak kind. And finally he understands why the mostly white non-military people living and working nearby were not pleased to see him walking among them. He is black.

Careful readers will have guessed this from subtle clues, but it is hard to believe that a black man from 1970 would not have anticipated and then recognized the race aspect of this brutal new world. After all, Martin Luther King was slain only two years earlier, and black fury had erupted against the white establishment. Perhaps as a highly educated and privileged black man Brian could close his eyes to these currents of bigotry and rage. But one suspects that Bob Tucker would not have predicted a president with a black father and white mother ascending before the close of the twentieth century.

Granted, it is a cardinal error to suppose that what happens in any given novel is entirely controlled by the political views and hopes of its author. Nobody would imagine that Wells really expected Martians to invade the Earth and lay it waste. But leaving that issue aside, Tucker's elegiac novel reminds us, again, that the primary role of time travel fiction is to confront radical physics with both inventive and pedestrian humans and watch what happens. As we have seen and will continue to see, wild fast paced action and adventure *is* often what happens. Not so in this case, or at least not as much as might be expected. If the novel ever has the misfortune to be made into a big budget movie, the title will probably be changed to *The Year of the Incredibly Loud Sun.*

8

Time's Up

1973 *The Man Who Folded Himself* David Gerrold

Best known (mostly to his chagrin, I imagine) for his invention of *Star Trek*'s furry tribbles,[1] David Gerrold (b. 1944) wrote what his publisher called "The Last Word in Time Machine Novels." There have been many additional sf novels and words about time machines in the subsequent 45 and more years, but that boast carried conviction in 1973 and to some extent still does.

For example, its gay writer might have been the first to acknowledge the possibility of a young time traveler, Dan, falling into enraptured physical love with his chronologically displaced self, Don. Later, in a variant on Heinlein's "Unmarried Mother" (from "—All You Zombies—"), Don finds his genetically tweaked female version, Diane, and inseminates her, fathering himself. Later still, as a very old man, he keeps a custodial eye on the boy until dying as a side effect of time travel shock.

Although marketed as a novel (and a finalist for the 1974 Hugo Award for best novel of the preceding year), this intriguing novella is a compact meditation on identity, maturation, love and its loss, and only tangentially on time travel, except as an enabling device. At his "Uncle Jim's" demise, Dan is bequeathed a complicated leather and metal belt that contains a miniaturized time machine and built-in set of instructions and controls. Where did it come from, who created it? We never gain a definitive answer, but Dan (or perhaps

---

[1] Readers and viewers might also recall his semi-autobiographical 2002 novel *The Martian Child: A Novel About a Single Father Adopting a Son*, based on his adoption of an autistic child and the boy's development under his care. It became the basis of the John Cusack movie *The Martian Child* in 2007.

© Springer Nature Switzerland AG 2019
D. Broderick, *The Time Machine Hypothesis*, Science and Fiction,
https://doi.org/10.1007/978-3-030-16178-1_8

Don, his twin other from one or more days ahead) does eventually consider several possibilities:

> …at some point, in some timeline, somebody invents a time machine. Somebody. Anybody…
> The first thing he'll do is excise the world in which the timebelt was invented, so no one else will… be able to come after him. Then he'll start playing around in time. He'll start rewriting his own life. He'll start creating new versions of himself; he'll start evolving himself across a variety of timelines. (95)

But perhaps the belt, once invented, is ruggedized, like his, for tyrannical military use. Even more unnerving, perhaps,

> somewhere there's a company that's manufacturing and selling timebelts like transistor radios. As far as each subjective traveler knows, he's rewriting all of time. It makes no difference… the number of alternate universes is infinite. (96)

Gerrold's reflections on the origins and utility of wearable time machines are thoughtful and provocative. In many time-traveling universes by other sf writers, difficulties with culture shock and unknown languages are solved by technologies that amount to magic by our standards. Download a cortical program that rewires your language centers and offers easy entry into social mores that have been shaped by hundreds or thousands of unfamiliar events and environments. By contrast, among numerous allohistories Dan finds a world where Jesus Christ and his followers failed to mark the Western Hemisphere with doctrines that in our world shifted human consciousness away from self and toward more expansive goals. That came with a price, alas: "Christianity has held back any *further* advances in human consciousness for the past thousand years" (94). Changing the past, almost by definition, cascades beyond calculation. It takes Dan years to understand this, and to abandon his selective killing of potential evil doers such as Hitler and Lee Harvey Oswald—even reinstating them back into life when the original intervention led to even more monstrous consequences.

In any event, this alienation effect, due to divergence between time lines, effectively limits Dan's access to most of the infinite variety of histories. The belt left to him by "Uncle Jim" has recognizable control abbreviations (F for forward in time, B for back, dates such as May and Mar) and help features in English. This suggests that the reality it came from was not far from Dan's own. Of course, who knows? Perhaps the device has an undisclosed ability to shape-shift from one cultural setting to another, and Dan simply lacks the knowledge of how to activate this convenient response.

As the book proceeds and his relationship with his other is replaced by a soaring, passionate, ebbing and finally defunct bond with Diane, Dan (or maybe Don) enters a world of multiples of himself, at different ages and interests. The young men are vigorous, devoted to demanding sports—hiking, swimming, motor sports, plus outdoor barbecues—and occasional group orgies. As they age, but continue rendezvousing in 1999 at a large mansion divided into rooms devoted to their disparate interests, the spacetime dynamics of this parable recede from sf into a kind of Robbe-Grillet "pure surface." At last the oldest of himself is desiccated, memory failing, and dies in the care of a middle-aged business-suited version of himself. Dan knows that this end will be his own, quite literally. Causality and sequence seem jumbled beyond the ability of a human mind to reconcile their threads, their beginnings, middles and ends, yet somehow Gerrold hints at a redemptive closure that is also an opening into unguessable possibilities to a man with a time machine wrapped around his waist:

> Uncle Jim has given his life back to himself—that is, to me. Now that I know the directions in which I shall go—no, *can* go—the decisions are mine…
>
> A whole world waits for me.
>
> The future beckons.
>
> Where to begin? (146–147)

1976 *Woman On the Edge of Time* Marge Piercy

At the outset, we proposed that a time machine "is a hypothetical device or condition of the cosmos able to propel a live passenger, or an instrument package, rapidly into the past or future. Preferably, it should be capable of returning that passenger or package to its time and place of departure." But suppose some people possess an ability akin to lucid dreaming, which is the well-established "altered state of consciousness" in which the dreamer is aware that she or he is indeed dreaming, and can to some extent modify the imaginary environment by just *wishing* it changed. The time traveling variety of lucidity might let you become aware in detail, as if you were physically there, of events in the past or future.

Would that be a kind of *traveling*, even though your body remained in the present? Could the biological brain and its manifestation as a *mind* reasonably be dubbed a *machine*? If so, one variety of time travel might not be detectably different from a surprisingly coherent hallucination. You might see and feel a futuristic or archaic setting, but not touch or influence it except by communicating with its inhabitants. The more one ponders this trope, the more plausible it seems that it would indeed be a bona fide kind of time travel. The self would thus conjoin both the neural bio-machinery and the phenomenological experiencer.

Woman On the Edge of Time implicitly makes this claim, while deliberately leaving undecided the reality status of these temporal excursions. The protagonist, 37 year old "fat Chicana" Consuelo (Connie) Ramos, born in 1938 (2 years after Piercy herself), is beaten brutally by her niece's pimp and trapped with broken teeth and ribs in a New York mental institution where the staff refuse to listen to her. Her purgatorial condition is Dantean: "Here she was with her life half spent, midway through her dark journey" (30). One dismissive question from a social worker captures the depersonalization to which she is condemned: "Where do you believe you feel pain?" (27).

Psycho-active and inappropriate medications of frightening crudity numb her mind and body. Connie experiences a pair of alternative tomorrows, Inferno versus Paradiso, somewhat akin to the bifurcation of possible futures in Jack Williamson's *The Legion of Time* and Poul Anderson's *The Corridors of Time* (1966). Even prior to her incarceration, she has communicated with the androgynous Luciente from the year 2137, whose confident demeanor initially misleads her. Luciente turns out to be a woman, with work-callused hands, from an anarchist or self-determining future culture where gender markers are elided in both speech and behavior: "he" and "she" become "per" (or person). Babies are gestated outside the womb and cared for by three parents of either sex; drugs make males capable of lactation. Wishful psychotic hallucination, or psychic time travel? Luciente explains that Connie is "an unusual person. Your mind is unusual. You're what we call a catcher... a person whose mind and nervous system are open, receptive, to an unusual extent" (41–42). Luciente perself is a sender.

Much of the novel recounts Connie's mental time travel journey through the community of Mattapoisett, learning the possible path of humankind out of her own victimized harrowing of hell, but seeing as well that alternative possible dystopian future. Toward the end of the novel, and under imminent threat of experimental brain surgery, Connie casts herself by error into the apartment of a young woman of the sexist, manipulative future such surgery foreshadows. There she startles and horrifies a young woman, Gildina

547-921-45-822-KBJ,[2] who sees her (so it is time travel, not delusion) and mocks Connie's lack of sexual appeal:

> [Gildina's] body seemed a cartoon of femininity, with a tiny waist, enormous sharp breasts that stuck out like the brassieres Connie herself had worn in the fifties—but the woman was not wearing a brassiere. Her stomach was flat but her hips and buttocks were oversized and audaciously curved. She looked as if she could hardly walk for the extravagance of her breasts and buttocks, her thighs that collided as she shuffled a few steps... small feet and tiny ankles and wrists... (288)

Introducing her fortieth anniversary edition of this powerful feminist novel, Marge Piercy stated:

> I projected a society [Mattapoisett] in which sex was available, accepted and non-hierarchical—and totally divorced from income, social status, power. No trophy wives, no closeting, no punishment or ostracism for preferring one kind of lover to another. No need to sell sex or buy it. No being stuck like my own mother in a loveless marriage to support yourself. In the dystopia [section], women are commodified, genetically modified and powerless.
>
> Before beginning the novel, I read all the utopian fiction I could lay my hands on, partly to study the narrative strategies that had worked and those that were too static to compel a contemporary reader. I also read at least as many dystopian novels—perhaps more...
>
> The other genre I was working in was time travel. I was weary of affluent white males hogging the genre.[3]

What establishes this narrative as a genuine work of time travel (despite there being no contrivance so vulgarly physical as a time *machine*) is the solidity and continuity of her experiences in the future/s. A somewhat similar tack would be adopted in Audrey Niffenegger's *The Time Traveler's Wife*, where both the husband and young daughter of the titular spouse find themselves flipping from *now* to *then* and even *thence*, propelled by a heritable DNA aberration. This is a storytelling device useful for reaching a broader readership than sf habitués and in particular one that regards itself as far more at

---

[2] Such alphanumeric names were an early prediction in sf, from Hugo Gernsback's 1925 serial *Ralph 124C 41+* to Ayn Rand's cast in *Anthem* (1938), such as Equality 7-2521, and Asimov's non-humans in "Nightfall" such as Theremon 762. On the cusp of the third decade of the twenty-first century, such apparent absurdities seem disturbingly predictive of gladly embraced hash (#) tags, Twitter and Facebook, and email identifiers such as B1lly54321@WTF.com

[3] See the edited extract at https://www.theguardian.com/books/2016/nov/29/woman-on-the-edge-of-time-40-years-on-hope-imagining-utopia-marge-piercy

home with what can be construed as metaphor rather than taken literally. Piercy herself has no such fear of generic contamination; a 1991 novel, *He, She and It* (aka *Body of Glass* in the British editions, and winner of the Arthur C. Clarke award), unreservedly deals with the inner life of an android cyborg, Yod, and its love affair with a human woman. She had already published a collection of sf short stories, *Small Changes* (1973).

Connie's political and personal radicalization impels her finally to deadly rebellion, a choice that derails development of the brain-remodeling mechanism that would have encouraged Gildina's detestable future rather than the liberative community of Mattapoisett. Her defiance, which requires the death of six people, is registered in history as a kind of Jonbar hinge, a moment critical in selecting one possible future rather than another (see the discussion earlier of Williamson's *The Legion of Time*). It is therefore not only Connie who travels in time; so too does her entire, collectively transformed heritage. In our real world, more than 40 years later, it seems all too likely that the resurgence of authoritarian, even neo-fascist regimes in the West, and the grotesque concentration of wealth and power in the hands of a minuscule proportion of the citizenry, is less likely to flower into Mattapoisett than Piercy hoped in the 1960s and 1970s. But her point in translating us into alternative futures is neither wishful thinking nor cynical despair. As she commented in her fourth decade Introduction:

> How is conflict dealt with? …who gets to decide, and upon whose head and back are those decisions visited? How does that society deal with loneliness and alienation? How does it deal with getting born, growing up and learning, having sex, making babies, becoming sick and healing, dying and being disposed of?…
>
> Utopia is born of the hunger for something better, but it relies on hope as the engine for imagining such a future. I wanted to take what I considered the most fruitful ideas of the various movements for social change and make them vivid and concrete—that was the real genesis of *Woman on the Edge of Time*.

1979 *Kindred* Octavia Butler

The most notable black American woman sf writer, Octavia Butler (1947–2006) is perhaps as famous for her 1995 MacArthur "genius award" as for her boldly conceived science fiction. Her time travel novel *Kindred*, unlike Marge Piercy's somewhat ambiguous *Woman on the Edge of Time*, leaves us in no doubt about the reality of Edana Franklin's repeated plunges into early nineteenth century Maryland. From Los Angeles to the antebellum horrors of a slave-owning and abusing plantation, 26 year old Dana is torn repeatedly from 1976 to save her white ancestor Rufus Weylin from near-death in 1815 and later (27).

These interludes when she is dragged away from her white husband, Kevin, display a temporal asymmetry. The first visit to the past lasts only seconds from Kevin's viewpoint, but minutes needed to save little Rufus from drowning in a river. In her second transition, the boy is now about nine years old and intent on setting fire to the heavy drapes in his bedroom, punishing his father, Tom, for beating him. Dana saves Rufus from burning the house down and perishing in the fire, and helps the boy fling the damaged, smoldering draperies into the night. When she returns to 1976, on the day of her birthday, Kevin reports seeing her vanish for less than three minutes, while for Dana hours had elapsed.

Time asymmetry poignantly afflicts Kevin as well. When she time traveled, Dana's clothing, complete with the switchblade that will set her free, remained with her. By the same token, on an occasion when Kevin was in physical contact her, he was also sucked into the past. Where, to the regret of them both, he was marooned for months when she was returned eventually to the twentieth century, obliged to flee to the north even though as a white man he was spared the utter hopelessness of a black person without papers. His letters to Dana addressed to the plantation were hidden from her; at last Kevin returned, profoundly altered by this immersion in an alien and frequently detestable culture.

A different species of asymmetry allows Rufus to perceive Dana when she is in her own time, although she does not share this kind of clairaudience. The boy says

I didn't see you, but I think I heard you.
   How? When?
   I don't know how. You weren't here. But when the fire started and I got so scared, I heard a voice, a man. He said, 'Dana?' Then he said 'Is it happening again?' And someone else—you—whispered, 'I think so.' (31).

Is Rufus, then, a receptive "catcher," with an ability akin to Luciente's in Piercy's sf novel, while arguably Dana is a "sender," not of transtemporal messages but of herself? That equation is too simplistic, though, because the woman is not consciously capable of reaching into the past; it is her youthful ancestor's desperate call for aid that triggers her temporal transition. And increasingly, she is trapped into remaining at the plantation by the young man's survival of various deadly threats. She frees herself by cutting her wrists, returns home, but reverts to the past once the wounds are healed Finally she escapes for good only because 25 year old Rufus threatens her with physical harm and the near certainty of rape, and she stabs him to death.

But they remain coupled in a gruesome way. His dying hand grips her left elbow, and Dana finds herself in her own house with that arm buried in the wall—an event portrayed in the novel's Prologue (9–11) as well as the close (260–261). The experience is excruciating, "an avalanche of pain, red impossible agony!" (261). We read this as an interpenetration of the atomic skeins of flesh and bone with the wall's plaster.

Well, then, is this not a variety of science fictional time travel, with the brain and body a machine working its will on the currents of time, and the physics of the external, non-personal universe wreaking damage upon the physicality of this mental machinery? Oddly, Butler denied that the novel was sf. In a 1990 phone interview with Randall Kenan, she observes that

> Most of what I do is science fiction. Some of the things I do are fantasy. *Kindred* is fantasy. I mean literally, it is fantasy. There is no science in *Kindred*. I mean, if I was told something was science fiction I would expect to find something dealing with science in it. For instance, *Wild Seed*... (Kenan, 495)[4]

Granted, there is as yet no *established* science of time travel and time machines, but neither was there such a background to many of the core themes and tropes enabling the development of classic sf. When Isaac Asimov set his galactic empire in a culture that moved between the stars via "Hyperspace" he drew upon certain speculations reaching back to the nineteenth century, but nobody was taking the idea of wormholes and faster than light transitions serious when he wrote these tales for *Astounding Science Fiction* in the forties and early fifties. It would be a cardinal error, though, to shelve Asimov's space fiction under the heading "Fantasy."

Gott and Thorne and other significant physicists and mathematicians, some of them Nobel laureates, provide us with hopes of time travel by various means, most of them fetched from afar in time and space but not thereby necessarily "far fetched." Additionally, as with the Piercy novel, drawing upon what amounts to the paranormal as a conduit for communicating and observing across the boundaries of time and space is indeed not yet agreed to be established science—but the laboratory and wild phenomena evidencing

---

[4] Randall Kenan, *Callaloo, A Journal of African Diaspora Arts and Letters* 14.2, 1991, 495–504; accessible at https://www.jstor.org/stable/2931654 I am grateful to Caribbean writer Gabrielle Bellot for this link and her perceptive Literary Hub article https://lithub.com/octavia-butler-the-brutalities-of-the-past-are-all-around-this/ October 17, 2017.

some kinds of *psi* effects are increasingly nearer to acceptance by brave scientists who have had the foresight to first gain tenure in their universities.[5]

It can be argued that the time travel in *Kindred*, and different mental powers in other Butler novels (such as *Wild Seed* and *Mind of My Mind*) are neither science fiction nor fantasy but rather metaphors for the act of imaginative engagement with the past or future. Certainly the emotional power of imagining a present-day person stepping into the racist past or future—as happens to a white historian in Ward Moore's *Bring the Jubilee*, to a black futurist in *The Year of the Quiet Sun* by Wilson Tucker, to Marge Piercy's suffering Latina in *Woman On the Edge of Time*, and to Butler's multi-generation *Kindred*—has necessary implications for us as empathic readers with a heart.

This claim, though, mistakes the mechanics of the narrative with its unfolding impact. Does considering the literary device of time travel in these harrowing books under the aspect of a *literal* or *mimetic* instead of *metaphoric* interpretation prevent or block a sensitive response? Surely not. However, it might be agreed that too narrow a focus on the hardware of an invented technology, if sustained exclusively, looks alarmingly like a disabling fetishism. Of course the mirror-reverse is no less unfortunate, as Emma Bovary found to her cost.

1980 *Thrice Upon a Time* James Hogan

Can we change the future? It seems boringly self-evident that we can and do, constantly—but then a moment's reflection suggests that if the future does not yet exist, we cannot *change* it. The best we can managed is to *shape* events and situations, to some extent at least, that have yet to occur. Still, at the very least it seems undeniable that we can intervene in the unfolding present to bring about consequences of that intervention that even *we* cannot foresee or modify in their entirety. A toss of a coin might seem to be a basic example, but actually it isn't. The unpredictability of a coin toss is really just a limitation of our knowledge of the determining forces at play as the coin springs into the air, tumbles too rapidly to be seen clearly by the naked eye, and strikes a surface that perturbs its final short path before it stops.

---

[5] See, for example, the four-volume set of now-declassified documentation from the two decades-long US government program on applied and theoretical paranormal capacities in Drs. Edwin May and Sonali Mawaha, eds, *The Star Gate Archives* (McFarland, 2018–2019). It is true, though, that so far such investigations have failed to transport humans to the past or future, nor to interpenetrate solid walls (luckily).

There is, though, a method that seems proof against any advance determination, and that is to make the choice crucially dependent on a quantum event. Quantum theory can provide us only with a probabilistic expectation—or rather, a whole range of them from highly likely to hardly likely—but Heisenberg showed why the final future event must always be to some extend indeterminate.

Well, if the future is always necessarily hazy and able to surprise us, what about changing the past, or even just helping to shake it? Surely the past is immutable? The much-loved *Back to the Future* movie trilogy made a meal of that proposal, showing aspects of the past such as a photograph literally dissolving in front of the camera and reshaping time's trajectory—but without expunging the memories of those who have returned from the future. In Part One we saw some analyses of such history-morphing, the kind investigated by Igor Novikov, Kip Thorne, Richard Gott, and other physicists. But usually their investigations are limited to extremely simple test cases, because anything more complex than a ball falling in and out of a looped wormhole are too mathematically taxing to get us very far. Science fiction, as we have seen, is rarely unsettled by the limitations of established science fact, and one method tested in sf does not require humans to travel physically back to the past in order to change the course of history. Sending a suitable message can do the job quite readily, as this novel by James P. Hogan (1941–2010) and the next one, by emeritus physics professor Gregory Benford (also born 1941), set out to prove, given their so-far fictional physics.

Hogan was British, involved with computer sales before taking up writing full time. This novel is set in Scotland, where Nobel laureate Sir Charles Ross theorizes about quarks and conducts experiments in his ancestral castle, Storbannon, as one does. In late 2009 his nephew Murdoch Ross, with an American friend, Lee Walker, are invited to the basement lab to view his time machine-in-progress. Here the three-quark components of protons are separated. (In the real world, this is forbidden because the quark confinement maintained inside the boundary of a nucleon *increases* as attempts are made to divide them.) Sir Charles claims that a new kind of radiation, tau waves, is revealed by quark separation which allows him to send messages, via a computer peripheral, up to ten minutes and ultimately a full day into the past.

Inevitably the young men hasten to create paradoxes to test the limits of this process. If they arrange to send a message back to themselves from ten minutes hence, and duly see it printed out *now*, they can in principle "bilk" the circuit by choosing instead *not* to send it back. How can this causality paradox be side-stepped? Perhaps by creating a branching parallel history, which is thereafter forever out of reach? At the outset, Murdoch speculates on "serial universes":

> Suppose that all the pasts that have ever existed, all the futures that will ever exist, are all just as real as the present. The present only gives the illusion of being more real because we happen to be perceiving it... (11)

This seems confirmed when a burst of conflicting phantom messages pour from the printer when the time machine's loop is bilked. Murdoch reflects that "future selves did exist who were just as 'real' as the selves that existed at a given present moment. That had to be so since somebody had sent the 'phantom' signals" (37)

Alternatively, perhaps the future selves who sent the message are *replaced* by the versions of themselves who have chosen differently in the present. So where did the print-out of the message come from? Why, from that now-inaccessible, or even deleted, segment of the future. Either way, mistakes made already can be corrected in advance—in the mistake's past. But if that obliterates the experienced reality of the future senders, why would they be motivated to arrange their own extinction?

We have to wonder, though, whether either of those pathways make logical sense. Isn't this just glib casuistry?

Hogan put his philosophical adventurers through several urgent and even poignant challenges that require one version or another of effective bilking. For the sake of exploring and dramatizing variants of retroactive messaging, Hogan piles disaster on disaster. Sir Charles has a friend, Elizabeth Muir, who works nearby at an experimental nuclear fusion plant, where by coincidence so does Anne Patterson, a lovely and intelligent physician with whom Murdoch falls in love after they meet by accident. To prevent a frightful problem—the creation inside the reactor of millions of micro-black holes that threaten the Earth's existence—Murdoch has to rewrite the history which brought him and Anne together. Other narratively convenient disasters are retrospectively thwarted, interfering with the past by bootstrapping a sequence of ever-earlier messages until crucial information is received in time for preventative and life-saving precautions (307). These options are accepted as trade-offs even though it means altering minor desirable events, such as Murdoch and Anne having met in the first place *and not even knowing it after the change.*

Sir Charles is aware of this dismaying kind of consequence:

It's futile to try and catch the system out by trying to set up paradoxes. All that will happen will be that the timeline will reconfigure to a new one on which all our memories and records... will be fully consistent with the new events resulting from the information pressed upon it.... It happens naturally all the time anyway [due to natural quantum fluctuations with unpredictable consequences]. Within the last few seconds you could have been reset from somebody who existed on another timeline to the one we're on at the moment, without ever knowing it. (221, 296)

Much of this action-movie-meets-doomed-romance scaffolding supports many pages of clever discussion of the theory and practice of time messaging. Some of the more canonical implications deployed by Hogan are pursued as well in Gregory Benford's rather more realistic and award-winning *Timescape*, to which we now turn.[6]

1980 *Timescape* Gregory Benford

A principle of sf creation espoused by plasma physicist/writer Gregory Benford is poet Robert Frost's "Play with the net up." That is, to the extent that it's possible, build your imaginary futures (or pasts, or sideways) in the context of what scientific theory and experiment have already established as valid or at least plausible, while accepting that one era's stern truth can be its successor's prat-fall. Phlogiston, the supposed fluid that explained the heat of combustion, gave way to the utterly different thermodynamic dance of colliding particles. Newton's universal time was slain by Einstein's variability of duration in different local frames. As the literary scholar Gary Wolfe put it in *Nature*,[7] sf authors who respect science

strive to incorporate actual developments in scientific theory into their fiction... Among these authors, the many-worlds hypothesis has joined the vast arsenal of shaky but convenient speculations—along with time travel, faster-than-light propulsion, uploadable minds, quantum computing and alien contact. These speculations, not yet fully testable in reality, provide continuing fodder for the ongoing dialogue between scientifically literate fiction writers and practising theorists. (2007)

---

[6] Published in the same year although an extract ("Cambridge, 1:58 A.M.") had been included five years earlier in the hefty original sf anthology *Epoch*, co-edited by Roger Elwood and Robert Silverberg.
[7] Gary Wolfe, NATURE|Vol 448|5 July 2007.

When James Hogan wished his researchers to send messages to the past, they built a machine that emitted and received "tau waves," which had the property of radiating backward in time. Such waves, suitably modulated, could transmit knowledge gained in the future into a past that might be altered by such information. That was an entertaining conceit, and propelled a novel which interestingly probed at classic and recent puzzles about causality and the inviolability of known facts about the past. But tau fields had the disadvantage of carrying with them a suspicion of narrative convenience.

When professor Benford set out in the mid 1970s to tell a somewhat similar time-messaging story, he chose to use a possibility that was already under consideration by theorists: the existence of faster than light particles, dubbed "tachyons," that never travelled *slower* than the speed of light in a vacuum. As we saw in Part One, it is demonstrable that in certain physical regimes this does not simply allow information, or even a starship, to zip from Earth to the far side of the galaxy very swiftly indeed, but to slide backward in time as well.

Of course, since this notion was first advanced seriously half a century ago, there has been a conspicuous absence of detectable tachyons in the world's most powerful labs (despite some early false positives). Searching for tachyons could be played with the net up, since its adherents were at least theoretically constrained by quantum theory and relativity. Sadly the game turned out to be played with an invisible and undetectable tennis ball.

None the less, Benford's use of the tachyon premise for fictional purposes gave his fine novel *Timescape* a solidity and rigor rarely seen before or since in time travel narratives.[8] (Ironically, Benford was one of the physicists who argued strongly that in reality tachyons could not exist. That did not prevent him from using the notion to underpin a novel rich in the prickly contests of the academy, and in the consequences for our planet's survival when political interests are pitted against uncomfortable claims—not unlike today's well-funded but empty assertions that anthropocentric global climate instability is "fake news," even when the vast majority of working specialists say otherwise.)

In 1998, the seas and oceans are clotted with fast-growing algae blooms driven by mutations in their genetic blueprint. Several lines of research converge on theoretical ways to have blocked this calamity before crucial ecological support systems were wrecked, placing humanity at risk of extinction—but it seems to be too late and the scale of the problem too immense for such intervention. While the problem could have been addressed effectively three or four decades earlier, the relevant research work had not been done at that time. Might it be feasible to find a way to send messages back to the early

---

[8] It won both the Nebula award and the John W. Campbell award in 1981.

1960s containing a solution to a catastrophe when such messages were not even detectable with available instruments?

Worse still, suppose a message *could* be sent and recognized (despite no dedicated receiver), decoded accurately, and used in the sixties and seventies to prevent doom. Would this success not subvert the whole causal chain of events via classic time machine paradox: if information from the future changes history retrospectively and saves the planet because the blooms never appear, then nobody would be motived to send such a message. If the scheme is to work, the message must provide only the barest hints of the issue and its solution, allowing experts decades earlier to make the necessary advances for themselves. Even then, the universe might simple bud off a fresh local version where the protected world is saved even though the original also persists and perishes horribly.

After some 40 years, Benford's novel retains its mainstream virtues alongside its undeniable value as net-up science fiction: a considerable amount of the characterization and settings draws closely on his own experience, a method of thickening the reality of the story dubbed "transrealism" by Rudy Rucker.[9] Its additional virtue is flagged by his Acknowledgments: "My aim has been to illuminate some outstanding philosophical difficulties in physics. If the reader emerges with the conviction that time represents a fundamental riddle in modern physics, this book will have served its purpose."

Curiously, one reason for the novel's longevity is that grand theory has stalled in physics after the consolidation now known as the Standard Model, with the Higgs particle its capstone and the failure, to date, to find extensions such as supersymmetric superpartners matching the known elementary fields and forces. And while tachyons have long been in eclipse, they have not been entirely ruled out. This allows non-scientists to enjoy Benford's articulate discourse on spacetime as understood in the mid- to late 1970s, especially the riddles of time that bear on the possibility of time machine messages.

At the University of California at La Jolla, in 1962, young physicist Gordon Bernstein and his graduate student Albert Cooper trace bursts of noise in a preparation of indium antimonide. At first this seems random gibberish, but they find recurrences when a signal seems to introduce a discernable spike into the trace. In 1998, at the University of Cambridge, British scientist John Renfrew and American Gregory Markham are responsible for these blurts 36 years earlier, as they attempt to transmit increasingly detailed messages in Morse code. An upper-class British overseer of this work and its funding, Ian

---

[9] This approach is discussed in detail in my *Transrealist Fiction: Writing in the Slipstream of Science* (Greenwood Press, 2000).

Peterson, finds their plan absurd but slowly comes to appreciate its potential, asking anyone in the past who detects their messages to leave a brief note in a La Jolla bank's safe-deposit box. When this confirmation is recovered, the Cambridge scientists are motivated to continue with their efforts despite savage cutbacks in research science. Both teams, despite the 36 years separating them, are repeatedly stymied by scoffers in positions of authority; Cooper fails his PhD while Gordon is denied an anticipated promotion. Meanwhile, the spousal, parental and sexually manipulative elements of recognizable life keep the narrative in focus as a mature history rather than a wild time opera.

One of the nicest touches in this novel of informational time travel is Benford's realization that signals backward in time, even when composed of collimated tachyons, have to take into account the orbital movements of Earth rotating both on its axis and around Sun, plus the Sun around the Milky Way galaxy, all of which combine to narrow the opportunities of a 1998 signal being detected at a certain point in space in the early 1960s. Clarifying this factor for any recipients, a repeated part of the early messages specifies where in the sky the transmitter will be, more than a third of a century into the future. This is given, at first somewhat bafflingly to these non-astronomers, as

RA 18 5 36 DEC 30 29.2

which Gordon guesses stand for the parameters Right Ascension and Declination, "*coordinates*, fixing a point in space" (158). With this hint in hand, they are able to direct their device to the most fruitful location, and finally receive details of the blooms and attendant viruses that will slay most people by the start of the next millennium.

Gregory Markham, seated high above a polluted ocean and a vast toxic cloud infected with the deadly virus, dies as the plane crashes, taking with it his brilliant breakthrough in understanding the impact a tachyon wave might make on the entire universe, rather like the "beep" in James Blish's *Quincunx of Time*. The trees beneath the doomed plane, he notices,

> rushed by faster and faster and Markham thought of a universe with one wave function, scattering into the new states of being as a paradox formed inside it like the kernel of an idea.—If the wave function did not collapse… Worlds lay ahead of him, and worlds lay behind. There was a sharp *crack* and he saw suddenly what should have been. (309)

With a detailed message specifying enough to allow the biologist Ramsey to synthesize a remedy for the algal bloom in its early stages, and a well-regarded woman scientist corroborating their work, Gordon's insight achieves recognition and the Enrico Fermi Prize after enormous resistance and mockery from his colleagues. He writes up a brief outline of his work for the magazine *Senior Scholastic*. A tall, beefy student, Bob Hayes, is sent by his teacher to the Student Book Depository in Dallas on November 22, 1963, and hears two shots. He finds Lee Harvey Oswald readying to fire a third round, and leaps, brings him down, saves President Kennedy's life. A new history calves off, free of the blight, an alternative trajectory through the timescape. Will tachyonic time machines entirely remake its future, entirely dissociated not only from the late Markham's ruined world but from ours as well? Perhaps so, but then we shall never know about it.

9

Highways to the End of Time

1982–2010 "Fire Watch" and Other Oxford Time Travel Works Connie Willis

"Fire Watch," sometimes given mistakenly as "Firewatch," won both Hugo and Nebula awards for best sf novelette of the year. The subsequent novel-length books set in the same time travel framework, but tonally diverse, have been equally successful among voting readers and critics. They are, to this date, *Doomsday Book* (1992), winner of the Hugo, Nebula and Locus awards, *To Say Nothing of the Dog* (1998), winner of the Hugo and Locus awards, *Blackout/All Clear* (2010), winner of the Hugo, Nebula and Locus awards. This impressively consistent achievement by Connie Willis (b. 1945) is a reflection of her adroit, emotionally engaging, deeply researched and often deadpan funny (even slapstick hilarious, as in *To Say Nothing of the Dog: or, How We Found the Bishop's Bird Stump at Last*). It is the time machine system that enables all this activity, joy and grief in ways that only temporal transition can permit, even when (as happens all too often and for a variety of reasons) it goes horribly wrong.

During the Nazi aerial attacks on England, one recurrent target of their bombing was the beautiful and historically important English Baroque cathedral, St. Paul's, built under the direction of Sir Christopher Wren (1642–1723) and seat of the Anglican Bishop of London. In 1940 it was saved repeatedly from conflagration by the astonishingly brave and dedicated rooftop patrols of a volunteer Fire Watch. In this novelette with that title, the cathedral was obliterated in 2007 by a terrorist pinpoint bomb, a source of persistent

© Springer Nature Switzerland AG 2019
D. Broderick, *The Time Machine Hypothesis*, Science and Fiction,
https://doi.org/10.1007/978-3-030-16178-1_9

anguish to the historians at the University of Oxford. Their experimental time travel unit serves as part of the educational process, inserting advanced graduate students undergoing their practicum—on-site work experience—into trouble spots, to observe and even participate. The most hazardous locales and dates carry a warning value of ten, and St. Paul's near-misses have been given that fearful evaluation. (The Blitz was only an eight.) Due to a computer glitch, John Bartholomew—who had trained for months in languages and other skills to observe Paul of Tarsus in the first century CE as he was inventing Christianity—is sent instead to St. Paul's cathedral while the incendiaries fall.

Bartholomew's roommate, Kivrin Engle, had already been stranded in the middle of the Black Death in 1349, after "slippage" in the time machine process—the Net—failed to drop her in comparatively safe 1320. Spared death by a suite of antibiotics and enhanced immunity, her harrowing in the depths of the Plague as almost everyone around her died horribly, from infection by a strain of influenza she carried back from the twenty-first century, has plunged her into what amounts to Post-Traumatic Stress Disorder. Kivrin's ordeal is dealt with powerfully in the first of the Oxford time travel novels, *Doomsday Book*, named for her electronic diary itself after the famous *Domesday Book*. In the novelette, returned from the fourteenth century, she plays a major role in saving her roommate and others from burning to death under the dome of the cathedral. Such repeated insertions of high-technology people into the past is at the heart of all these Willis explorations of early experimental time travel by the academy. Her distinctive approach blends old-fashioned adventure narrative entertainment and poignancy, sometimes heartbreaking. As diseases she might have cured, with drugs she does not possess, slaughter the villagers she has come to love, especially Agnes, a young girl, she denounces the monstrous medieval God of her time-lost friends:

> You bastard! I will not let you take her. She's only a child. But that's your specialty, isn't it? …
> Kivrin washed her little body, which was nearly covered with purplish-blue bruises. Where Eliwys had held her hand, the skin was completely black. She looked like she had been beaten. As she has been, Kivrin thought, beaten and tortured. And murdered. The slaughter of the innocents. (380)

The startling shift in *To Say Nothing of the Dog* is toward light-hearted whimsy, especially rollicking confusion largely due to the neurological side effects of "time lag." These are akin to the drastic disruption of sleep associated with intercontinental jet lag, but deranging perceptions. It is odd that this analogy has rarely been seen in earlier time machine or teleportation fiction, but it is perfectly reasonable to expect that abrupt changes in level of ambient light and other environmental factors would cause just such mental blurriness and perhaps even hallucinations; "one of the first symptoms of time-lag is a tendency to maudlin sentimentality, like an Irishman in his cups or a Victorian poet cold-sober."

Further tonal delights come from Willis's invocation of other texts, most obviously the titular borrowing from Jerome K. Jerome's famous comic novel *Three Men in a Boat*. Other additives are drawn from P.G. Wodehouse's fictional butler Jeeves, a servant of exquisite sense and practical competence, bumptious Oxford professors with outlandish views, and the loud American Lady Schrapnell, an abominable manipulator whose quest for the hideous bird stump is a major driver of the story. And of course the bulldog Cyril.

Primarily this is a love story between historian Ned Henry and Verity Kindle, whose repeated bungled attempts to heal ruptures in spacetime find them stranded in different eras. A cat carried into the future by Verity threatens mayhem in a history where all felines have perished from distemper. Certain of the laws of temporal physics are disclosed: travelers apparently cannot re-enter times they have previously visited, while incongruities might change the course of history, or even delete the universe. Luckily, this does not happen, unlike the denouement for one probability universe in Michael Swanwick's *Bones of the Earth* (discussed below).

The most recent Oxford novel, too large to be published profitably in a single volume, appeared as a diptych: *Blackout*, set in 1939 and 1940, and its sequel *All Clear*. Both are organized, as with "Fire Watch," around the horrors of Nazi attacks on Britain and the resulting enormous social disruption. In each of these books and the original novelette, History Fellow and time project researcher James Dunworthy, now in 2060, plays a key if bumbling role. Plotting of the diptych, all 1100 pages of it, is richly complicated, much of this confusing elaboration due to effects of time travel. In addition, historians sent to observe the Blitz operate under different names at different times.

Pretty Polly Churchill presents as Polly Sebastian, a shopgirl, after having been Mary Kent, a nurse, in an earlier visit to the 1940s. But that "early" visit is calibrated to her personal history, and actually occurred some years previously. Unless she can find a drop zone allowed her to escape back to 2060, the time physics law forbidding two versions of the same person co-existing implies that she will die shortly unless she manages to get back to the future before 1943. This sort of tangle engulfs most of the traveling historians. Oxford student Michael Davies adopts the name Mike Davis, ends up by error on a boat evacuating soldiers from Dunkirk, is badly wounded but terrified that his involvement in this critical event might have derailed history in favor of the Germans. Merope Ward travels to 1939 as servant Eileen O'Reilly in an aristocratic country house that is subsequently commandeered for war work, including supervision of young children moved to safely from bomb-torn London. Here she takes in hand siblings Alf and Binnie Hodbin, who catch measles and might have perished if not for Merope's commonplace medical knowledge from later in the century.

This interweaving of past, shifting "present" and future leaves one of the historians in post-war Britain, of her own choice, while others are led to a drop zone by a teen boy obsessed with time travel and Polly. An understanding of this cat's cradle of timelines and protective duplicity is reached, suggesting that some slippage results from time healing itself. If these time travelers are prevented for an interval from going home, it is because their presence in the past allows their participation, however apparently insignificant, in bringing about victory over the Nazi forces. Time travel, on this account, cannot bring about the best of all possible worlds, but it can spare us from the worst.

1985 *A Maggot* John Fowles

Marvelous in both senses of the word—beautifully written and crafted, and resonant with marvels—*A Maggot* is perhaps even better than Fowles' *The French Lieutenant's Woman*. That novel (if not the variant movie spun from it) was an attempt at a splintering multiverse love story, although few of its admirers would have made that sf connection back in 1969. Even fewer recognized that *A Maggot* is a time travel narrative, however much it echoes with deceptions and yearnings and blocked sexuality and a utopian longing for God and community.

John Fowles[1] (1926–2005) made an early commercial success in the mode of postmodernist uncertainties, although he repudiated that description as modish in another sense.[2] On the face of it, this masterpiece from Fowles in his late fifties seems a blend of convincingly wrought historical invention and the dawning of a religious impulse in a young woman, Rebecca Lee, trapped by her eighteenth century patriarchal constriction in a life of prostitution. It is all that, but secretly and emergently renders a hypothesis closely aligned with the theme of this book: a time machine hypothesis for the emergence of

---

[1] His name is pronounced "Foals."

[2] "Convinced of 'the intrinsic slipperiness and ambiguity of words,' he was still not ready to accept the 'fetish-disciplines' of postmodernism" (Eileen Warburton *John Fowles: A Life in Two Worlds*, Viking: 2004, 453).

a faith community, This becomes the humble, egalitarian and all-but-celibate Shakers under the guidance of Ann Lee, the real daughter of the novel's fictional visionary on a Pauline journey to a new trust and hope.

Six people, Lee the only woman, set out in 1736 from London on a kind of occulted pilgrimage from, or perhaps to, a heart of darkness in the rural South West English country of Devon. All six are to some degree imposters. They travel under the direction of a young aristocratic Lord, Mr. Bartholomew (or Mr. B.), disguised son of a powerful and wealthy Duke. Mr. B. seeks "the secrets of the world to come—I mean not those of heaven, but of this world we live in... what shall happen tomorrow... a thousand years from now.... Further suppose that this prophet reveals that the predestinate future of this world is full of fire and plague, of civil commotion, of endless calamity" (23–24) Bartholomew poses this to Mr. Brown, actually Francis Lacy, who is presented to those they meet along the way as the younger man's uncle and master of the undertaking but is in fact an actor. Mr. B. is also attended by a deaf and dumb man, Dick Thurlow, with whom he grew up. Thurlow—whose surname, we notice, seems lexically rather too close to "[Bar]*tholo*[me]*w*" for mere chance—dies en route, hanged apparently by his own hand, perhaps a Judas demise. David Jones, or Sergeant Farthing, plays a soldier who protects the small group from the hands of thieves, ruffians and bothersome divines.

What is their goal? Several inconsistent tales are offered, but it seems that when Mr. B., Thurlow and the woman enter a cave at the end of their journey, this has been known by Bartholomew as his destination. Perhaps he has been here previously; at any rate, he is not surprised to find a large "maggot"-like vessel sequestered somehow, impossibly, inside the dark inner portions of the cavern. Something disturbing happens there—an attack by demons or bandits? a visitation from unknown realms? something too wonderful to disclose to the unredeemed? Different accounts are offered by Lee, one suggesting diabolic witchcraft or the like, the other a far more sui generis and baffling recital to the Duke's legal investigator. Henry Ayscough is a small man with a sharp, intensely conservative but not unimaginative mind. He cannot fathom this absurd tale from the quiet young prostitute, but *we* can, because the advances in technology since the eighteenth century have been so immense that we know at once that this "maggot" is a sophisticated transport of some kind.

But what kind? Many commentators seem to have decided that it is a spaceship from another world, carrying Lee into the dark beyond the sky to witness their peaceful, sexually segregated alien culture. However, to a science fiction reader, it is immediately apparent that this device is a time machine—

capable of flight, no doubt, and able to teleport in and out of the cavern, but with a crew of fully human if transcendent visitors from the future. They bear witness to Lee of a different way of life, one that would eventually find practical expression in the cult fashioned by her daughter Ann, "the United Society of Believers in Christ's Second Appearing, better known as the Shakers."[3]

Oddly, Fowles began in the expectation that this instrument of illumination would be a spaceship, come to rescue survivors from an impending nuclear apocalypse. As his biographer Eileen Warburton recorded in 2004, Fowles started but abandoned an sf novel 30 years earlier:

> His notes give various titles: *The Mother Planet, The Final Account, A Last Account, The Survivor, Adam and Lupela,* and, at last, *The Screw.* In the fictional fragments, Earth is completely devastated in a nuclear war in the year 2075. The narrator, a forty-five-year-old writer, is saved just before the catastrophe by the female voyagers of a spaceship from the planet Lupela. To the "Ladies, infinitely strange in form, infinitely kind of heart, my saviours," the earthman tries to explain his culture, the intricacies of human language, and the destruction of nature on his dying planet. The spaceship, "spherical… stationary and silent," and its wise female travelers are forerunner of the vessel and voyagers of the supernatural vision in *A Maggot.* Fowles was imaginatively at work on that book more than a decade before publishing it. (361)

But of course the vessel visited by Lee was neither "a supernatural vision" nor, primary, a spaceship. We are shown with precise clarity that it was a time machine. Nor, it seems likely, was her experience within the control room a veridical flight into space or across the sky to somewhere else on Earth hidden from everyone, but rather a holographic travelogue movie on a large display, rather didactic in format and perhaps intended more as indoctrination than as candid reportage. Perhaps, though, that suspicion might just be cynicism aroused by our own far from perfect social orders.

Here is Rebecca's claim: they are met as they approach the cavern by a woman in plain silver clothes, including narrow trousers and black leather boots. His Lordship takes a respectful knee and doffs his hat, and Dick and Lee do the same out of baffled courtesy. She has dark hair and eyes and olive skin. At the entrance to the cavern Lee bathes and drinks, noticing a newly burned circle on the ground (something that UFO claimants often mention). Within, she sees something floating, "like a great swollen maggot, white as snow upon the air" (359): an antigravity machine? Its white hull resembled "fresh-tinned metal, large as three coaches end to end, or more, its head with

---

the eye larger still… other eyes along its sides that shone also, tho' less, through a greenish glass. And at its end there were four great funnels black as pitch, so it might vent its belly forth there" (359–360). It hangs in the air, its circumference ten or twelve feet. A spoked wheel is painted in blue upon its side, and other icons, one the now familiar white-and-black taijitu or Tao symbol (hardly likely to be known to a harlot of the city, nor to drop-in space men and women).

She detects a purring hum, and a sweet odor. The maggot floats down with a sigh and stands on thin legs with large "paws." A door is opened, and latticed stairs lead from the opening to the ground. The first woman is joined by one older and one younger, and the younger two merge into the older woman. She places her arms around Mr. B. like a mother embracing a long lost child, and they speak together in an unknown tongue. He casts aside his sword, a shocking thing for a British Lord to do in that time. Rebecca, his Lordship and Dick are invited into the craft, where she sees arrays of illuminated "precious stones" marked with signs she could not read but that we recognize from any electronic instrument panel. The ceiling lights are replaced by darkness, and they see through the small clear windows followed by "a greater prodigy than all, for where was the chamber's end, that stood before the maggot's head, was of a sudden a window upon a great city we glid above, as a bird" (370–371), some three feet by four. But they are not in reality flying, for she astutely notices that "I might see by those smaller windows we moved not from the cavern, its walls still stood outside" (371). What the "window" shows is not continuous but evidence of editing: "here, from afar, here close," and so on (372),

There follows enraptured descriptions of a city of white and gold but set in ample gardens, orchards and fishponds, with great golden highways where people traveled in horseless carriages, apparently propelled by motion in the pavement of the street. No churches are evident. "And the sun shone on all, like to June eternal. So now do I call this happy land that we were shown" (373). There were no children "of the flesh," since the sins of the flesh were absent. She describes the Shaker ideal life, as preached in years that follow by her as-yet-unborn daughter.

Now the ontology changes, grows mystical. Two men were one, father and son, "our Lord Jesus Christ"; the woman is not Mary the virgin but the Holy Spirit, Holy Mother Wisdom (378–379). The investigator is upset by these heresies. The tone darkens. Visions of cruelty and desolation are shown, murderous warfare. By no coincidence, these hellish sights are just what the three children claimed to witness in Fátima, Portugal, on October 13, 1917 (see discussion in Part Three), shown to them by the young woman they conclude must be Mary and who displayed her credentials in a fulfilled prophecy of a falling and cavorting device of light that the thousands of witnesses took to be the Sun "dancing."[4] And the Lord, Mr. B., vanishes, returned to June Eternal (385). What this theophany is meant to convey is perhaps open to every reader, but the Lordship's behavior seems closer to the mischief of Loki than to one of the holy Apostles of the New Testament or the prophets of the Old. We might conclude, after this remarkable journey into sacred terror and wish fulfilment, that time machines, like Trojan Horses, are not necessarily to be trusted when they come calling.

1986 *Highway of Eternity* Clifford Simak

Third author chosen by the SFWA for the honor of Grand Master of the genre (in 1977), pastoral poet of early twentieth century Midwestern values, Clifford Simak (1904–1988) published this frustrating wander through the landscapes of eternity two years before his death. Writing well and thoughtfully in one's eighties is not unknown, but there is usually a sense of dislocation and weariness in such work. Oddly enough, Simak's fiction at the midcentury height of his powers already had a rather grandfatherly tone, like the meditative mood music of many of his recognizable characters each watching the grassy landscape from his stoop, perhaps with a friendly robot at hand. Time under such an ambience of dying falls is often distorted in his fiction, from the early novel *Time and Again* (1951) to his Hugo-winner *Time Is the Simplest Thing* (1961) and in this final volume more than 35 years later.

As science fictional explorations of time travel (or, in some cases, interuniversal travel), this body of work is unsatisfactory even when actual time machines are invoked, such as the devices—machines called "travelers"—in this last novel. It is not that Simak failed to explain the working parts and purpose of the traveler-devices. If you were asked to transport a family group of Cro-Magnons in your horse and carriage, it might be irritating to be nagged with clueless queries about the biology of this animal and the purpose of the harness. What's frustrating about *Highway of Eternity* is the blurry lack of context. When war journalist hero Tom Boone is revealed as a man who can "walk around a corner" if threatened with extreme harm, perhaps passing for a short time in a higher dimensional realm, this is not intrinsically problematic for sf. When his old CIA operative friend Jay Corcoran turns out to have a gift for seeing deeper into the *Maya* of the world, after major head damage

---

[4] See, for example, Joaquim Fernandes and Fina d'Armada, *Celestial Secrets: The Hidden History of the Fátima Incident*, Anomalist Books, 2007.

in a plane crash, that is rather an odd coincidence, Still, maybe people with superpowers tend to stumble over each other and become chums.

These two, seeking missing Corcoran client Andrew Martin, zip to Shropshire in the year 1745 using a traveler-device disguised as Martin's small invisible balcony apartment stuck on the outside of a tall hotel about to be demolished. Here is a place named Hopkins Acre, vanished from the world in 1615, and they find in a manor house under a force field-protective bubble a time traveling family from around the Year 975,000. When we learn that their names are… David, Emma, Enid, Horace and Timothy, not to mention their semi-ghostly discarnate brother Henry, we are irritatingly trapped in the aspect of the novel called by critic John Clute "a good deal of sheer silliness" (*Science Fiction Encyclopedia*, Simak entry). It's possible, admittedly, that their names are really Ruby 124C 41 and Gzopp Pimw, and so on, but Tom and Jay show no curiosity about this until David a million years later explains the general principle to Corcoran who asks:

> To you I should seem a shambling, uncouth primitive; to me you should seem a sleek sophisticate. But neither of us finds the other strange. What goes on? Didn't the human race develop in all those million years?
>   You must take into account that my kind were back-country people," David said. "The hillbillies of our time. We clung desperately to the old values and the old way of life. Perhaps we overdid it, for we did it as a protest and might have gone overboard (117)

He adds that his far future home time was emptily utopian, with disease defeated, plentiful sleep, food and medicines, and by the way "The human lifetime has been more than doubled since your time" (117). The sheer failure of imagination in this feeble extrapolation can't be readily explained by the expectation gap between now and the early 1980s. The reason for this meager lifespan advance derives from the engine of the plot: aliens, the "Infinites," have abandoned the flesh (long an sf trope, this) and now exist as immortal minds without the limitations, passions and trivial concerns of organic humans. They have made it their task to convert everyone on Earth to pure incorporeal energy and information.

To escape this allegedly gruesome fate (which, when you think about it, resembles the blessed postmortem paradise of major religions), rebellious refugees such as the Evans family and others have scattered across the checkerboard of the past. On the run again now, after an assault by the Infinites on the Hopkins Acre homestead, the Evans family plus Tom and Jay run into more than a few colorful or at least unusual human and nonhuman entities.

These include a noble wolf who becomes Tom's friend in an American desert setting 50,000 years ago; a beseeching but murderous robotic web that Tom disables with the rifle David usually liked to carry everywhere unloaded; a many-eyed critter they name Horseface who with bookish Enid's help (or so it seems) is building a better spacetime traveler; robots who make it their role to cut down and burn every tree on the planet, because trees might otherwise become humanity's heirs; and The Hat, a mysterious figure who seems to have escaped *avant la letter* from the Harry Potter universe and is a sockpuppet representing an unknown player (or so it seems).

Despite this inanity, notable sf writer Michael Swanwick cites two brief but striking passages in the novel that

> evoke that most hoary of science fiction virtues—the sense of wonder. A little of which can make up for a great deal of what otherwise was a terrible waste of time… The plot is a rambling, arbitrary mess. Multiple suspensions of disbelief are required to keep it going. The implications of the enabling technologies are pretty much ignored….[5]

As the narrative bounds, or in some cases limps sluggishly, toward a conclusion where explanations are multiplied and tested to destruction, secrets and role reversals are exposed like carved horses breaking free from a calliope and running off in all directions. The Highway of Eternity is a gray, featureless zone pierced by a white line that leads to the Galactic Center where many intelligent aliens gather to ponder the mysteries of the universe. The key to unraveling these mysteries might be Tom Boone, even though the assembled magi of philosophy have been gnawing at these puzzles for a million years or more. "We are thinkers and investigators," an alien explains. "We try to make some sense of the universe" (215). Like a version of *Dune* sketched by E. Nesbit, Boone (with his unparalleled quirk with stepping around a corner) and Enid are brought together so that their unusual gifts will merge to provide answers to these vexing questions:

> From their union, there was a chance that a new race would spring—an offshoot of humanity that combined the evolutionary trend shown by Boone and the toughness of that small group of humans who had stubbornly dared to resist the menace of the Infinites. (287)

---

[5] Swanwick, http://floggingbabel.blogspot.com/2016/12/forgotten-sf-clifford-d-simaks-highway.html

All of this hodge-podge is laced together by time machines of varying heritage and capacity, stolen, lost, forgotten, recovered. An entity in the Galactic Center campus tells Timothy, "We stole the process for constructing the machines from the Infinites. We had no part of developing time travel. We simply blindly followed a pattern. We knew almost nothing of the technology" (221). This disclosure is falsified for the reader but not the principals when an apparently buffoonish alien reflects with some satisfaction that he had supplied the Evans family and others "with one of the most simple of the time machines, developed by his race as an ancient forerunner of the net. The Infinites had time travel, of course, but theirs were such complicated devices that the rebels could not have understood them. Lying again…" (287).

And one of those primitive time machines, and the surviving malign AI core of the web destroyed by a bull bison in the Pleistocene desert, allowed Andrew Martin to escape to the twenty-third century and launch a gaudy religious cult that led eventually to the near-vanishment of embodied *Homo sapiens*, transformed into sparkles of pure mentation. The lesson of the novel: In this lapsarian world, you really have to keep your eye on sf's time machines— almost anything can happen if they fall into the wrong hands.

1989 *"Great Work of Time"* John Crowley

In the premier closing position of his annual *Year's Best* science fiction anthology for 1990,[6] the late Gardner Dozois introduced this brilliant long novella as "intricate, subtle, luminous, and mysteriously evocative," and it is all of that. John Crowley (b. 1942) had emerged as a sophisticated and ingenious writer of sf and fantasy novels and short fiction in the mid to late 1970s, and created a magical world of the fantastic in 1981s *Little, Big*. "Great Work of Time," published for the first time in his collection *Novelty*, is a tour de force in the themes of time travel and its relationship to ideology, dreams, wishes, and fate.

A cross-genre gem, it won the 1990 World Fantasy best novella award, and can be read that way, but really it is time travel science fiction even though it has no machine in its working parts, other than consciousness and the seething inaccessible machinations behind our awareness of self and world. The secret of this method of visiting and manipulating the past is discovered by the idiosyncratic Caspar Last in 1983 and used to take him to the previous

---

[6] This anthology was published in the USA as *The Year's Best Science Fiction: Seventh Annual Collection*, but in the UK as *Best New SF 4*, both released in 1990; it is the latter edition I cite.

century when the British Empire was, in effect, the world. Last desires neither power not excessive wealth, only "a nice bit of change" to let him pursue his abstractions. Once he's made half a million or so, he plans to destroy the documents of his "time machine," based on orthogonal engineering. "Caspar always thought of his 'time machine' thus, with scare-quotes around it, since it was not really a machine, and Caspar did not believe in time" (532).

It does, however, succeed in manifesting his eidolon, with a beard abruptly down to his waist, beneath a British Guiana plantain tree in 1856. Purchasing an envelope, he addresses it to his great-grandfather and posts it with a one-penny magenta stamp affixed. Returning to 1983, he finds it among his ancestor's papers, and prepares to sell it to the consortium that owns the only other surviving stamp of its kind, now worth a million dollars. They will set it on fire, he expects, after paying him half a million, funding the remainder of his life in first class.

But at this point, Caspar effectively leaves the story, shifting to upper-middle class and Oxford-educated Denys Winterset, born in 1933 and now, in 1956, on his way by train to Southern Rhodesia. A former assistant district commissioner in Bechuanaland (now, in our history, the Republic of Botswana) in the British Empire's Colonial Civil Service, he pauses at Khartoum on his way to his new posting and is drawn into conversation by Sir Geoffrey Davenant, an adventurer operating for the Colonial Office or some more rarefied station.

Davenant proves to be a member of the Otherhood (nothing so wimpy as a "Brotherhood") and has sought out young Denys in rather the way, it turns out eventually, that Asimov's Harlan was drawn into Eternity because his pivotal role in its long-ago establishment was known from secret history records. His own crucial role, it turns out, is to visit Cecil Rhodes in the past, in territory that was now Rhodesia, and murder him before he grows old and changes the will in which he left his immense fortune to the Otherhood. Denys is aghast, naturally, but the necessity of this single intervention in time is brought home to him. Unless Rhodes dies at the height of his power, a different future will emerge without the benefit of Otherhood's redemptive step by step improvement and sanitation of the Empire.

In fact, what follows is a brutal history we recognize as our own: two monstrous global wars, nuclear weapons, the near genocide of the Jews by Hitler, and abominable colonial policies crushing the rightful aspirations of people of color. (Small arcane jests are scattered throughout the novella, as is often the case with Crowley's work. One, surely not by accident, is Davenant's name. In the seventeenth century, Sir William Davenant [1606–1668] was an innovative Royalist Poet Laureate, playwright, godson of William Shakespeare and perhaps named for him, who regarded himself as "Shakespeare's Heir"; one of his plays was *The Siege of Rhodes*.)

This crux is where we, and Denys, come to understand that no history is dominant, but neither is the actual dynamic one of shunting from a given thread of the multiverse to another causally dependent on the induced change (such as shooting Rhodes dead). In Denys's experience, in 1893 he draws the Webley revolver, and in the lawn of the *kraal*,

> seeming in that illusory light to be but a long leap away, a male lion stood unmoving… I could see the dim figure of a gamekeeper in a wide-awake hat, carrying a rifle, and Negroes with nets and poles…
>
> I was sure, instantly sure, that a lion which had escaped from Rhodes's lion house had appeared on the lawn… just at the moment when I tried, but could not bring myself, to murder Cecil Rhodes…
>
> And I know that in fact there was no lion house… Rhodes wanted one, and it was planned, but it was never built. (585–586)

So Denys, or fate (whatever that might mean), had conjured the lion and its residence out of his unconscious and into reality, like a dream or reprieved nightmare solidified. From that moment, Denys Winterset moves or is moved from one alternative landscape and set of inhabitants to another and yet another, finding himself in a world where a Magus (a nonhuman species, it seems) catechizes him in the true large reality, a world of angels, highest of the hominids:

> "And how many times since then," the Magus said, "has the world branched? How many times has it been bent double, and broken? A thousand times, ten thousand? Each time growing smaller, having to be packed into less space, curling into itself like a snail's shell; growing ever weaker as the changes multiply, and more liable to failure of its fabric; how many times?" (570)

None of these deleterious changes are planned or intended; they result from the exponential accumulation of small errors and divergences, beyond computation, perhaps admixed with the dream seeds like Rhodes's lion. Finally, in Winterset's perhaps insane and delusional perception, the final state of the world is without bird, beast, fish, or humankind, Magi, draconid servants, Sylfids, nor Angels. All the world is sunk into near immobility, the forest in the seas (531), immense trees rooted into the mud of the sea depths, leaves moving silently in the blue currents. It is the future of almost terminal entropy. This, then, is the meaning of the title, drawn from Andrew Marvell's poem about Cromwell:

To Ruin the Great Work of Time
   & Cast the Kingdoms old
   Into another mold

A molded reshaped architecture of spacetime and its infinite orthogonal rotations, reduced, inevitably, perhaps into the slime of mold. But even this dire entropic reading is put in question by Crowley's commentator, who notes that the society of the Otherhood are limited in their yearning for a perfect universe, leading therefore to the silent trees in a soulless cosmos. It is "we, out here, who live in but one of innumerable possible worlds" (591) who endure calamities but choose ways of confronting them other than the reductive fates of Caspar Last's discovery, at last.

10

Windows Into the Past

1991 *Time's Arrow* Martin Amis

In 1917, the German baby Odilo Unverdorben ("Uncorrupted") died in the brutal depths of the First World War and was absorbed into his mother's uterus. Later, he became John Young, then Hamilton de Souza, and finally, absurdly, was born as Tod T. Friendly after a heart attack. *Todt*, of course, is German for *Death*. Unlike every other human on Earth Tod T. lived his life in reverse, observed as he did so from within—including shared dreams but not thoughts—by a kind of endlessly confused but chirpy guardian angel or soul or reified narrative viewpoint, a singular chorus to a life of genocidal depravity. Even this unusual summary is not quite correct, as we shall see.

Here we have that extremely rare variety of time travel, a consciousness with its arrow of lived time twisted one hundred and eighty degrees. Martin Amis (b. 1949) is certainly one of the most brilliant and daring novelists of the last half century, and no other literary icon had adopted this dismaying, harrowing and appallingly funny story line. But Amis did not originate negative chronology. F. Scott Fitzgerald (1896–1940) published "The Curious Case of Benjamin Button" in 1922, following the biological regression of its titular character from birth in 1860 as a shrunken oldster with the power of speech to his death 70 years later as an infant. He aged backwards, but his awareness of himself and the world remained locked to the passage of time shared by the rest of the world.

In the strictly science fictional mode, retrograde life was explored at length in Philip K. Dick's novel *Counter-Clock World* (1967), which Amis *fils* must surely have read; his father, Sir Kingsley Amis, was a notable sf enthusiast,

© Springer Nature Switzerland AG 2019
D. Broderick, *The Time Machine Hypothesis*, Science and Fiction,
https://doi.org/10.1007/978-3-030-16178-1_10

critic (*New Maps of Hell*, 1960), anthologist and occasional contributor. In Dick's rather baggy novel, the Hobart Phase brings everyone back from the grave, although not simultaneously. We are told the resurrected experience time in reverse but this seems mainly confined to metabolic activities. Nutrition, for shock-value example, is supplied via an anal nozzle, reconstituted internally, and reconstructed food items are ejected in due course from the mouth.

The same horrid process seems to be at work in *Time's Arrow* as well, but reported with careful indirectness: "Later, you pull up your pants and wait for the pain to go away. The pain, perhaps, of the whole transaction, the whole dependency… it seems to me like a hell of a way to live" (11). What we witness here is not time itself running backward, any more than it reverses for Benjamin Button; it is rather the experience of the narrator-entity that is being rewound from end to start. When the birds sing strangely, when everyone walks backward, he asks "What is the—what is the sequence of the journey I'm on?" (6). It is only his temporal viewpoint that reversed at the moment of death, and what this immaterial soul sees and feels seems to be a form of postmortem "life review," as proposed by those who believe in survival of death. This point is never made explicitly in the novel, but what afflicts Friendly on his journey toward birth and childhood as young Unverdorben might be the terrifying fate of each of us, awaiting judgment on our life choices.

On the other hand, perhaps even on this reading there is no judgment from a metaphysical beyond, only an eternal Nietzchean return from A to Z then back to A then back to Z then back to A then… forever. This strikes us as preposterous and unphysical, especially in a universe with quantum indeterminacy at its root. Even so, some physicists continue to support a macrocosmic version of such a cycle. The conformal cyclical cosmology advanced by Oxford University's Sir Roger Penrose suggests that "the universe cycles from one aeon to the next, each time starting out infinitely small and ultra-smooth before expanding and generating clumps of matter. That matter eventually gets sucked up by supermassive black holes, which over the very long term disappear by continuously emitting Hawking radiation. This process restores uniformity and sets the stage for the next Big Bang."[1] That model does not require each cycle to be a mirror-image of its predecessor, however. Rather the smoothing aspect, even if there is no obliterating singularity, would seem to prohibit any retention of prior structure or memory (as exemplified in Poul

---

[1] https://physicsworld.com/a/new-evidence-for-cyclic-universe-claimed-by-roger-penrose-and-colleagues/ (August 21, 2018).

Anderson's "Flight to Forever"), perhaps even of the fundamental parameters of each emergent aeon.

Playing with such cosmogonic speculations in the context of the appalling evil of the Holocaust has seemed to some readers not merely insensitive and shamefully frivolous but wicked. While it was shortlisted for the Booker in 1991, a number of reviewers denounced Amis's novel in those terms, before it became more fashionable to dismiss it as simply arch, "the type of clever trick that the *Twilight Zone* used to toss off in a half hour," and "the time reversal technique is not compelling enough to carry the story along for all of its hundreds of pages."[2] (Actually the Vintage edition runs to 165 pages, not much longer than a novella; its brevity contributes to its whiplash impact.) In this study of traveling in time, the imagined redemption by time reversal of millions of dreadful murders and medical atrocities, focuses our attention on the real world's evasions, rewriting and erasure of memory, mob embrace of enormities.

The use by Amis of reverse causality jolts us awake again and again. Dr. Friendly's deteriorated relationship with "a really big old broad. Irene… as she opened her legs I was flooded with thoughts and feelings I'd never had before. To do with power" (35, 37). For much of the narrative, Tod is impotent, although as he grows into his repaired bodily strength and attractiveness he managed to abandon and seduce any number of nurses. His marriage with Herta, who visits him in Auschwitz in 1944, ("the gaping universe of the Kat-Zet," the Zone of Interest or concentration camp) is ruinously cruel; their daughter dies in infancy.

What is especially nauseating is his soul's misunderstanding of his toil. Starved, naked bald women patients have their shaved hair replaced, the Jews are gifted with clothing, "to prevent needless suffering, the dental work was usually completed while the patients were not yet alive" (121). He has been funded in the earlier/later years by his share of this extracted gold. As they leave the Kat-Zet, the prisoners are given a ring or other small valuable. In general, this is the nature of his reality: "I look up and form the hilarious suspicion that the world will soon start making sense" (106). It never does. He works to removed pellets of Zyklon B from the dead, passing them to the pharmacist. Pimps remove the bruises and heal the cuts of their prostitutes with a slap of a bejeweled fist. Infants are squeezed into the womb with forceps. Smoke particles fall upon the crematoria from the sky, bringing new Jews to life:

---

[2] http://brothersjudd.com/index.cfm/fuseaction/reviews.detail/book_id/361/Time%27s%20Arrow.htm

Creation is easy. Also ugly…. Here there is no why. Here there is no when, no how, no where. Our preternatural purpose? To dream a race. To make a people from the weather. From thunder and from lightning. With gas, with electricity, with shit, with fire. (120)

Nearly at the end, after these grotesque and salvific explanations, a chapter opens: "Well, how do you follow that? The answer is: you can't. Of course you can't" (137). After a Holocaust, there is only silence. And yet, no, that is not true. Memory is, indeed, a kind of time machine, and we keep ourselves sane by renewing recollection. One day, perhaps, we shall have machines allowing us to view such monstrosities and even, perhaps, visit them against the drag of time's arrow, if not undo their maleficence. But for now, we have fiction as well as testimonials from the aged and the latest victims from some new hell-hole on the face of the planet.

1991—*Outlander* Series Diana Gabaldon

One variety of temporal transition we have not yet examined might be dubbed "time slippage," where some segment of space become detached from its time order and appears as a portal into the past or future. This has become almost a sub-genre in its own right, and can appeal to readers with no appetite for science fiction of the "with rivets" kind. So it is not really *time machine* fiction, because it is usually beyond the power of reason to explain or understand. Borrowing from medical terminology, its process is a kind of *temporal anastomosis*—a cutting of one portion of an era from its previous causal strands and its re-stitching into another frame of reference. This can result from huge tectonic shudders and breaches in the structure of spacetime, as in Sir Fred Hoyle's *October the First is Too Late* (1966), Gordon Dickson's *Time Storm* (1977), and Arthur C. Clarke's and Stephen Baxter's Discontinuity in *Time's Eye*, *Sunstorm* and *Firstborn*, together forming *A Time Odyssey* (2003, 2005, 2007).

The most successful sequence in this mode, although without the global catastrophic side effects, is probably Diana Gabaldon's much-loved nine book romantic *Outlander* series (1991 to forthcoming), and *Outlander*'s so far four-season TV incarnation (2014—). In this narrative of cross-century transitions, the time slip is closer to sorcery than anything explicable by science. The franchise has proved more lucrative than any other time slip fiction (selling over twenty-five million copies of the book sequence, and making Gabaldon the 14th most influential woman on a 2014 *Hollywood Reporter* list)[3], presumably because it is *not* regarded by its fans as science fictional, and does not draw upon any sort of constructed time machine.

In 1945, former British Army nurse (and orphan) Claire Randall visits Scotland with her husband Frank and suffers a time slip into 1743. She and Frank had visited a standing stone circle and found there fifteen local women, young and old, performing a kind of unclad Druid ritual. The following drizzly morning, seeking certain rare plants, Claire returns to the tallest stone, where its vertical slit had been drawn open two or three feet. The vulval imagery is deliberately ambiguous, implying both coital promise—the previous night, Frank had undressed her as they lay in the grass, "siding his hand under my skirt and up my thigh to the soft, unprotected warmth between my legs" (33)—and the wailing of traumatic, re-born parturition as she passes dizzyingly through the cleft:

There was a deep humming noise…
    The stone screamed…
    I had never heard such a sound from anything living… It was horrible.
    The other stones began to shout. There was a noise of battle and the cries of dying men and shattered horses. My vision began to blur.
    I do not know now whether I went toward the cleft in the main stone, or whether it was accidental, a blind drifting through the fog of noise….
    I was quite sure I was still hallucinating when the sound of shots was followed by the appearance of five or six men dressed in red coats and knee breeches, waving muskets. (34–36)

The man in charge proves to be Frank's cruel and sexually perverse ancestor, Captain Jonathan "Black Jack" Randall, who closely (and thus genomically) resembles her husband. Clearly this affinity-by-marriage is part of the motive force and temporal directedness that drew Claire into a key place and time of her historian husband's ancestry, the spooky stone circle location which had seen similar vanishments. An sf reader might wonder immediately if her childhood orphan status suggests that she was smuggled as an infant 200 years forward into the future, and is now drawn back to her true homeland. This guess is undermined, but perhaps not definitively disproved, when it turns out that she was reportedly the daughter, born in 1918, of Julia and Henry Beauchamp, a couple who were burned to death in a car crash when she was five.

---

[3] http://www.scotlandnow.dailyrecord.co.uk/lifestyle/outlander-creator-diana-gabaldon-named-4731629

Threatened by Black Jack with rape, Clair is saved by a rebel against the British Crown who takes her on horseback to a stone cottage where are gathered a rough bunch, including a badly wounded hero whose shoulder has been torn out of its socket. Claire corrects their clumsy resetting efforts, cleans the wound, and in the course of the novel falls in love with this very masculine young Highland warrior, Jamie Fraser, the fugitive Laird of Lallybroch. Under pressure she marries him to gain Scottish status, in a sort of time-inverted bigamy. She loses her first baby, Faith, later becomes pregnant with Brianna and returns to the twentieth century, to everyone's astonishment and suspicion, not least her long abandoned husband's, for the difficult, life-threatening delivery. The multi-novel sequence, with further temporal transitions, is immensely detailed with plenty of surprises (Claire has sex with the King of France, for example) and emotionally involving, but this is not the place to elaborate or synopsize those many aspects of the books.

In any event, we may ask: how does time slip operate in Gabaldon's universe? Specific locality—standing stones, for example—is not the only factor. A time traveler gene is involved (as in the 2003 Audrey Niffenegger novel *The Time Traveler's Wife*), evidenced by heritability of the gift, which usually only gets activated in dire circumstances.[4] Precious stones are also involved. So science, of a sort, is not entirely excluded from what might otherwise seem occult superstition.[5] At least one young woman she meets, Geillis Duncan, regarded as a witch by the yokels, is a time traveler from the 1960s. Claire's daughter and Brianna's husband Roger McKenzie "begin to record everything they know—or at least suspect—about time travel in a journal, which they jokingly refer to as the 'Hitchhiker's Guide' for time travelers. The journal includes exposition on ley lines and geomagnetism as part of a series of working hypotheses on how time travel works."[6] One might be tempted to doubt the validity these findings, but Brianna has a degree from MIT in mechanical engineering and has the mind of an artistic scientist:

> Slowly, as it always did, the calm inexorable logic of the figures built its web inside her head, trapping all the random thoughts, wrapping the distracting emotions up in silken threads like so many flies. Round the central axis of the

[4] Gabaldon has commented: "Well, see, nobody who has the time-travel gene is going to *know* they have it, unless they fortuitously find themselves in rather rare circumstances." (Sept 12, 2009, https://web.archive.org/web/20160410181359/http://forums.compuserve.com/discussions/Books_and_Writers_Community/_/_/ws-books/65160.12?nav=messages).

[5] See details at https://outlander.fandom.com/wiki/Time_Travel

[6] See "A Practical Guide to Time Travelers" at the site above.

problem, logic spun her web, orderly and beautiful as an orb-weaver's jeweled confection. (*Drums of Autumn,* 1996)

1997 *In the Garden of Iden* [The Company Sequence] Kage Baker

> Once there was a cabal of merchants and scientists whose purpose was to make money and improve the lot of humankind. They invented Time Travel and Immortality… In reality it was the other way around. The process for Immortality was developed first. In order to test it, they had to invent Time Travel. (*Iden*, 1)

When the brilliant Elizabethan theater expert Kage Baker (1952–2010) died of cancer at age 57, she had been publishing science fiction for less than a decade and a half, but her impact was swift and enduring. Her major work was the Company/Dr. Zeus, Inc. series of nine novels and a considerable number of short stories and novellas linked by the theme of time travel, covert manipulation of history and prehistory, and enforced immortality for a few.[7] Happily for her enchanted readers, Baker was able to complete this significant million-word-plus storyline with *The Sons of Heaven* (2008) and *The Empress of Mars* (2009), which satisfactorily and quirkily tied together the multitude of threads spun out of the fertile premise of *In the Garden of Iden*. That debut novel introduced her obsessed and time-harried Spanish botanist Mendoza, its narrator, seized as a small girl and imprisoned in 1541 by the Inquisition, saved by a Company Facilitator, Joseph, and augmented into a deathless cyborg by dubious emissaries from the twenty-fourth century.

Baker's approach to time travel, now often advanced by specialists in the dynamics of closed timelike curves, is that history (at any given moment) is invariant. Whatever is known to have happened, and reliably recorded, cannot be changed by visitors from the future. That is not as much a restriction as one might suppose, given that much history is unrecorded or open to interpretation. Thus, time travelers might intervene covertly.

> If history states that John Jones won a million dollars in the lottery on a certain day in the past, you can't go back there and win the lottery instead. But you can make sure that John Jones is an agent of yours, who will purchase the winning ticket on that day and dutifully invest the proceeds for you. (2–3)

---

[7] A useful, detailed account of the Dr. Zeus, Inc. project and its many operatives and conspiracies is http://en.wikipedia.org/wiki/Dr._Zeus_Inc. but this should not be consulted before reading the novels.

Centuries later, wisely husbanded by financial dealers throughout the past, your winnings will arrive in your present in the form of funds, land, or recovered "lost" artworks by the now-famous—lost only in the sense that they have been sequestered by representatives of the Company. Even extinct creatures and plants can be preserved for the benefit of future ages, so long as you have reliable agents seeded across the ages. This scheme allows the Company access to the treasures of time even though nobody can travel forward beyond their own point of origin. (Or so it seems, until Mendoza puzzles scientists with a distinctive spacetime-warping "Crome radiation.")

Cyborged immortality, needed by these Company proxies effectively stranded in the dark ages of history and pre-history, proves workable only when massive structural and biological changes are made as early as possible. Adult bodies and brains, it turns out, are too settled to allow such radical improvements. Doomed, forgotten children of the past are located by Company agents and press-ganged into eternity. Feisty, ignorant little Mendoza, even her name borrowed, is found condemned to death as a Jew by equally ignorant but far more culpable Inquisitors. Rescued by 20 millennia-old Joseph, she begins her transformation and accelerated training as a specialist in rare plants for her saviors from nearly 1000 years in her future.

By the mid twenty-fourth century, this furtive organization—sometimes known as Dr. Zeus, Inc.—has in effect become rulers of the world. But an increasing number of indications suggest that its future will not be the utopia that its immortalized delegates have hoped to embrace, with their manumission, at the end of their long, weary journey. The evidence is indirect, but powerful: no messages have been received from beyond the Silence in July 8, 2355. Might the world end on that day, perhaps in a global conflict brought on by the Company itself? It is a fertile narrative premise, and Baker needed nine novels to complete her journey into time: *In the Garden of Iden* (1997), *Sky Coyote* (1999), *Mendoza in Hollywood* (2000), *The Graveyard Game* (2001), *The Life of the World to Come* (2004), *The Children of the Company* (2005), *The Machine's Child* (2006), *The Sons of Heaven* (2007), *and The Empress of Mars* (2009).

A master of story, she was now emotionally moving, even heartbreaking, now zany and laugh-out-loud black-humored. Mendoza's fate, eerily, is to fall desperately in love three times with the same cyborged man, who is different each time and always inappropriate, but compulsively desirable. It is the kind of tangled gothic romance only possible in a time travel sequence, and Baker worked it for all it is worth, to our enjoyment and benefit. In the opening volume, he is a sixteenth century English puritan, the former libertine and radical Nicholas Harpole, racked with religious guilt over his sexual infatuation with the lovely, mysterious Mendoza, who is only just starting to get the smallest notion of what being a cyborg implies.

In England for the marriage of Queen Mary Tudor to Prince Philip of Spain, Mendoza visits the botanical garden of Sir Walter Iden and finds there a rare medicinal treasure, Julius Caesar's Holly. Hence, the novel's title, with its inevitable undertones of exile from the Garden of Eden and its secret trees of both life and the knowledge of good and evil. Under Joseph's cynical tutelage, Mendoza seduces Nicholas, but when he uncovers her inhuman nature he flees to Rochester and preaches fire and brimstone, attracting the wrath of the reinstated Catholic church. Like the famous Oxford Martyrs, he is sentenced to burning at the stake. White-skinned, red-haired, black-eyed Mendoza, with her enhanced powers and knowledge of futuristic technology, struggles to save her lover.

While the opening volume of the sequence remains very much the best place to start the saga, it really comes into its own with Sky Coyote. Mendoza arrives on the west coast of America in 1699, before the ruinous arrival of the Spanish. The local people, the Chumash tribe, are taken in by Joseph's manic impersonation of a trickster god, but less persuaded by his efforts to recruit their guilds into service of Dr. Zeus's Company. Accomplished in dialect, Baker avoids the usual stilted translations, presenting the Native Americans in charming and sometimes hilarious colloquialisms crossed with wheeler-dealer Chamber of Commerce-speak from three centuries later. Introducing Joseph as he visits Humashup township, chief Sepawit stands in "nothing but a belt and some shell-bead money":

"Well, folks. I guess our distinguished visitor doesn't need much of an introduction to you all—" Scattered nervous giggles at that...

"Uncle Sky Coyote, I'd like to introduce Nutku, spokesman for the Canoemakers' Union...

And this is Sawlawlan, spokesman for the United Workers in Steatite." Another one wearing lots of money, with big hair and a sea-otter cape. "And Kupiuc, spokesman for the Intertribal Trade Council and Second Functionary of the Humashup Lodge. And this is Kaxiwalic, one of our most successful independent entrepreneurs." (Sky Coyote, 82–83)

There is not a breath of condescension or mockery here. Baker shows us worlds ancient and recent as if we live in them, and for all their startling familiarity they remain hauntingly strange. This is wonderful time travel storytelling, cut tragically short by Baker's early death.

2000 *The Light of Other Days* Arthur Clarke and Stephen Baxter

Introducing the discussion (above) of T.L. Sherred's 1947 *Astounding* story "E for Effort," I noted that a temporal mechanism displaying the past in images and maybe sound is fundamentally a more plausible prospect than time machines carrying a traveler into the past. In essence, a time viewer is no more absurd than watching a movie made 50 years ago, or gazing at a distant battle through binoculars. You can't change the past events displayed that way, however much you might wish to, and however terrible the consequences might be of proving that the saints and gods of all religions are inventions and that the causes of righteous wars were often grotesquely concocted (as historians often reveal, but never so convincingly as a time viewer peering into the past).

Perhaps the most comprehensive survey in fictional form of the effects of time viewing of the past, with the future remaining inaccessible, is this collaboration by Sir Arthur C. Clarke (1917–2008) and Stephen Baxter (b. 1957), with its titled borrowed from a poem with that title by Thomas Moore (1779–1852). Acknowledging this source in their Afterword, Clarke and Baxter note as well that it was used in 1996 (without "The") to title a much loved *Analog* tale by Bob Shaw.

Actually that version was far more limited, for its supporting trope was "slow glass," a transparent material that slowed the transmission of light by a very significant factor, allowing people in weeks, months or years to witness scenes it had captured in their past. Once the light waves had completed their transmission from the back of the glass, each frame of the image was gone, since the slow glass was only a valve rather than a storage medium. This ingenious narrative device made a kind of sense, since light is indeed slowed as it passes from vacuum to air to water and thence eventually to laboratories where even more ingenious scientists have managed to slow light almost to a standstill in the decades since. Sadly, Shaw's device would produce cloudy images as the incoming waves interfered. What's worse, the accumulated light and heat inside the pane would eventually shatter the ever-hotter glass. Imagine how much energy would fall on a window if the sunlight from an entire year were trapped within it. Shaw accepted this critique and modified his imaginary mechanism in later slow glass tales, but the first fine simple beauty of the idea was lost.

Clarke and Baxter used a more sophisticated method, based on relativistic and quantum research ideas discussed in Part I, above: wormholes, where spacetime is warped so that a voyage from one mouth to the other might be minutes, years, even millennia in duration for an outside observer in one framework while far briefer for the traveler in an accelerated frame. Indeed, as we saw in the examples of the paradoxical billiard balls passing backward through time and perhaps slamming into themselves at an earlier point in their original trajectory, quantum tunneling can in principle permit and enable visual images and other information to be ported from the past to the future. If an extremely tiny quantum wormhole could be fished out of the foam that underlies our macroscopic reality, it might be expanded and reshaped to create such a one-way aperture, and perhaps guided hither and yon anywhere in time and space, yielding more and more knowledge of people and events not only from familiar history but also from the ages of dinosaurs, to the earliest living cells, as well as the secret doings of princes, priests and paupers, and the neighbors next door.

In 2033, a bright journalist, Kate Manzoni, releases prohibited news of a very massive comet or similar space body, on a course to impact the Earth in some 500 years, perhaps as soon as 300. Nicknamed Wormwood (from the Book of Revelation), it is forty times the mass of the dinosaur killer 65 million years ago, and will strike the planet with immensely more force. All life, perhaps, will be scoured from the surface of the planet one hundred and sixty feet deep. The convulsive psychic impact of this news disrupts the work and attitudes of the world, even though everyone alive now will be long dead.

Why struggle to correct global climate? For what future generation? Cultists create virtual reality experiences of heaven, such as RevelationLand, making billions from the dread of billions. Already fantastically wealthy, Hiram Patterson (born Hirdamani Patel, after his family fled from Uganda to England) and his handsome son Bobby now have a research facility near Seattle, the Wormworks. With the help of his French-schooled mathematical physicist son David Curzon, captive wormholes are developed as undetectable cameras, enabling reporters such as Kate to have inexpensive, intimate and instant access to trouble spots and secret policy cabals all around the planet.

As the novel proceeds, staying close to these three and the sociopathic Hiram, breakthroughs from the Wormworks repeatedly rewrite the understanding of current physics but also expand the technological impact of these findings. At length, once WormCams become commercially available to everyone, cultural changes occur at every scale, without the fatal nuclear consequences imagined by Sherred in "E for Effort." The result is not unlike what is happening now with cellphones in almost every pocket or handbag, ready to photograph criminal behavior by high and low alike. One of the intriguing aspects of this book written some two decades ago is the extent to which the authors correctly forecast the impact of computerized media (Google,

Facebook, Netflix, etc.), driverless vehicles, the appeal of extreme religious consolations even as young people increasingly abandon prudish constraints and embrace decent acceptance of gay, trans and other minority ways of life.

One factor explored in different ways by Baxter in other novels is the emergence of the Joined, a kind of hive mind of individuals who remain individuals while co-experiencing with others via brain-connecting technology. At death, these shared repositories will sustain the consciousness of the physically defunct (probably), and ultimately permit resurrection even of the long dead into cloned bodies build from WormCam-scanned DNA, the new humankind's long-term goal.

More immediately, what happens when WormCams are devised that reach into the past and find that Moses was a composite of numerous leaders and currents of belief, as was Jesus (in terms of miracles, virgin birth, and being God)? Interestingly, the co-authors tread carefully here (they bypass Islam entirely), and reveal that when viewers try to look at the Crucifixion of Christ everything goes blurry:

> WormCam exploration there is limited. Some scientists have speculated that there is such a density of viewpoints in those key seconds that the fabric of spacetime itself is being damaged by wormhole intrusions. And these viewpoints are *presumably sent down by observers from our own future*—or perhaps from a multiplicity of possible futures... (220)

Perhaps the most satisfying and sustained imaginative feat directly driven by the time viewer is a joint investigation of David's timeline, following the path of his mother's, grandmother's mitochondrial DNA through hundreds, thousands of generations. Because the mitochondria pass only from mother to daughter and son alike, the female line is never broken. All the way to the birth of organic life. At that point, the WormCam viewpoint can be diverted to the unfolding retro history of the planet, from centuries of drought to centuries and more of glacial ice, from evolved competences of creatures to their extinction in some local cosmic catastrophe—all watched in reverse, to the dawn of the Earth as a world under an increasingly looming moon seen through a methane atmosphere. There are two concerted chapters on this theme, remarkably detailed and affecting. If we ever develop time viewing prowess, even if it only reaches into the past, experiences such as this will surely have a profound effect on humanity. Luckily, Clarke and Baxter offer us a literary shortcut to that dazing experience. (By the way, SPOILER, you will be relieved to learn that Wormwood is brushed aside a century hence, like a moth—a technophilic outcome that Sir Arthur would surely have demanded.)

11

From Dinosaurs to Elsewhen

2002 *Bones of the Earth* Michael Swanwick

In 2010, an Administrative Officer named H. Jamison Griffin brings a cooler to the Smithsonian Museum office of paleontologist Richard Lester, and offers him a job but no details other than an impossible lure: the cold but recently living severed head of a Stegosaurus. Lester, astonished, spends the next twelve hours dissecting the dinosaur's brain. It is totally convincing, not a fake. Could it be a *Jurassic Park*-style DNA composite? No, far too difficult. It has to be a specimen brought to the twenty-first century from the Cretaceous via time machine. A year and a half later Griffin invites him to join the classified government time travel project.

Michael Swanwick (b. 1950) is one of the brilliant sf and fantasy writers of the class of 1980, with a dazzling firecracker career since his first story, "Ginungagap," was published that year not in *Analog* or *F&SF* but in the biannual literary magazine *TriQuarterly* alongside one of the most remarkable stories by Thomas M. Disch. His work, like Disch's, combines masterful style, political acumen, impressive ingenuity and an undercurrent of sophisticated humor. *Bones of the Earth* has the added pleasure, for those who like this sort of thing, of doctoral-level familiarity with the ecology and multiple long-extinct life-forms of the deep past, deployed with effortless specificity. (A much shorter version was published as "Scherzo with Tyrannosaur" and won a Hugo award in 2000).

Swanwick's theory of time travel in this novel suffuses the narrative but largely without any attempt to explain its operations, although its source is finally revealed. The familiar paradoxes are discussed and generally dismissed

© Springer Nature Switzerland AG 2019
D. Broderick, *The Time Machine Hypothesis*, Science and Fiction,
https://doi.org/10.1007/978-3-030-16178-1_11

as unphysical, yet closed temporal loops are commonplace but constrained by one fundamental Genesis-scale rule: Do not do X (for various kinds of significant causal violations) or the project will be retrospectively obliterated by the mysterious deep future beings who provided the time machine technology. "…any threat to this precious and fragile enterprise will be nipped in the bud… And those responsible will be punished. No exceptions…" (25) Worse, the entire time line might be erased:

> What if, in spite of your best efforts, a paradox slips by you?…
> Cut free from causality, our entire history from that moment onward would become a timelike loop and dissolve. (25–26)

Might this explain the extreme paucity of time tourists in our known history? Perhaps. But merely apparent paradoxes are not forbidden. Griffin gathers potential travelers together in a ballroom and demonstrates the basics, preparing to leave the podium and travel for half an hour by limo to the Pentagon, where he will travel an hour into the past, return by limo to the Grand Ballroom and shake hands with the Griffin who is addressing them now before leaving the podium. This happens, but the mechanics remain unexplained. Clearly, this brief doubling is not itself a paradox, merely a tuck in spacetime with no causal violations. Later, we witness Griffin sending reports back to his earlier self that are used to prevent severe harm to the project although not always to its participants. A kind of doctrine of predestination constrains effective actions—until, finally, it fails and the consequences are very serious indeed.

The secrecy of time travel is abandoned on July 17, 2034, and three generations of paleo and other scientists gather and visit each others' stations, careful not to drop hints of the future to those from their past. One of these experts is Gertrude Salley, an ambitious and charismatic woman who trashed Leyster's work when she was 25, but continued romantic or at least highly sexualized dealings with both him and Griffin. Salley, as she preferred to be called when young, is the scientist who announced the reality of time travel to the press. The news enrages cultists, especially the deep creationists who advocate violence in defense of their religious lunacy. Robo Boy is a warrior of the Thrice-Born Brotherhood, deliberately trained in those sciences and technologies that ensure he joins a survey team to the Mesozoic, 225 million years before the present. He plants explosives in the time beacon needed for retrieval of the team. It does not kill Salley because Griffin, under instruction from the Old Man, himself many decades older, chooses to spare her life. Immense consequences follow.

For most of the events in the novel, this paradox infraction is background only, allowing readers the pleasure of viewing the rise and collapse of dinosaurs and other beasts through various Periods, Epochs and Ages. In a collective version of Robinson Crusoe, a team led by Lester is marooned in the aftermath of the beacon explosion, and in the spirit of their devotion to science they build shelters, learn how to fish and hunt, observe many forms of life's intersections, and develop wonderful hypotheses.

Some dinosaurs sing, others use near-infrasound messaging to herd lesser or stupider creatures. One scientist proposes an impressive model for the great dying that follow the Chicxulub impact that wiped out the dinosaurs around 65 million years ago. The slow, subtle grinding of the tectonic plates, he suggests, provide a pulsed background infrasound cycle guiding the herder animals in sustaining layer upon layer of ecological order. When the giant rock slams in from space, the Earth rings like a bell, disordering the long-established pacemaker for centuries, and everything falls apart for the great beasts, freeing niches for mammals and other small animals. (Presumably this notion is Swanwick's own devising, and it is exactly the sort of novelty that makes good sf such a delight for the intelligent reader, even though it almost certainly cannot be true.)

By the close of the novel, these effective adventure tales and more paleontological naming and observations than you can shake a spear at (including a spear wielded by a woman that actually slays a T. rex) are revealed as two distinct history-variant universes of a Many Worlds variety but shaped to the wishes of very far future humanoids. These are the Unchanging, a stern monastic species of identical individuals who carried the science of time travel back from 49.6 million years hence (it takes megayears to invent time travel), and beyond them a terrifying non-human species, evolved wingless birds in the single coalesced continent Ultima Pangaea. They have studied humankind, delighting in our beauty and nobility even as the line holding the time travel project is negated. This bold narrative move risks suspicion of being a shaggy god joke, but Swanwick snatches us from this tangled time loop to biographies of the survivors from Richard Lester's team, one by one, movingly because we have now lived among them and watched them grow together. The purpose of time travel, we can feel truly for a moment, is indeed to both display and witness the nobility, decency and beauty even of flawed and primitive beings like ourselves.

2003 *The Time Traveler's Wife* Audrey Niffenegger

In the first two decades of the twenty-first century, the fiction of time travel once again came to the fore, enlivened particularly by work from women writers who revived and reshaped some of the now-routine tropes. Some of these women writers were consciously aware of the sf megatext, that enormous collaborative distillation of ideas and treatments including artificial minds, exploration and command of space, non-human aliens, and post-Wellsian voyages to past, future and alternative histories. Joanna Russ, Connie Willis and other expert writers, well-armed with this toolkit and adding to it, had already opened sf to the many voices of women. Now a new expansion occurred, as "mainstream" writers adopted versions of the time machine hypothesis, often without nuts and bolts but propelled by genomic changes, say, or unexplained rewinding and modified replaying of events.

One of the first works in this rekindling was *The Time Traveler's Wife*, a best-selling debut novel by Audrey Niffenegger (b. 1963). It is a powerfully realist novel grounded in the premise of moving convulsively and without choice back and forth in time, fantastika enriched by copious details drawn from Niffenegger's real life. Certain telling aspects of the world are familiar from literary fiction but not always included in commercial sf. What distinguishes *The Time Traveler's Wife* from, say, a clever traditional sf entertainment like Poul Anderson's *There Will Be Time* (1972) is its very ordinariness in the midst of the utterly unsettling. It refuses, by and large, to use this intrusion of an sf *novum* as an opportunity to showcase the time traveler's technical prowess, political *nous* or prejudice, trans-historic destiny.

Here is a revealing quote from the writer:

> I like science fiction, but it's not really what I read. So I wasn't trying for science fiction... what I was initially interested in was having one fantastical or strange thing and then regular reality. There's this idea that you change one thing about the world and everything else moves around it. This idea that you're allowed to play with reality somewhat. In my art, I'm somewhat surrealistic.... I like changing things.[1]

Sf, of course, is all about changing things. What happens when you are telling a contemporary story—even one where a 6-year old girl is likely to be

---

[1] Veronica Bond, interview with Niffenegger, Bookslut, 2003: http://www.bookslut.com/features/2003_12_001158.php

visited by her forty-something, stark-naked future husband—and an
overwhelming world-historical event intrudes into your own life, into the
narrative arc of your book? "The part that happened around 9/11 was interesting," Niffenegger told an interviewer,

> because, of course that happened when I was almost done with the book and I
> thought, wow, I can't really let this go un-addressed. For the most part real
> world events don't really make it into this book because I didn't want to date it
> and I didn't want it to be about the world. It's really about this relationship. I
> figured, you have this gigantic thing and if you don't at least nod at it, it's going
> to seem glaring in its absence.

This is a version of time travel fiction of the quotidian, in this case amplified into gritty terror that serves in the narrative, for a brief moment, as an
icon of the traveler's uncanny, dreadful, fated, powerless standpoint, ever
moving, never moving. Niffenegger comments, "It's something that bugs me
about actual science fiction, this effort to provide all the answers and make
everything work out very neatly." But of course her novel *is* "actual science
fiction," at least if Daniel Keyes' much-loved *Flowers for Algernon* is sf. (In that
novel, neurosurgery carries a likeable retarded man to genius and is then powerless to prevent his slide back to his initial condition.) Still, Niffenegger is
right to feel qualms; her novel is this kind of increasingly popular sf more
often written by women than men, enlivened and enriched by familiarity
used against itself to provide a jolt not only of shocked surprise but also, paradoxically, of recognition.

Both are features of her emotionally moving time travel novel. Librarian
Henry DeTamble and his once and future wife Clare Anne Abshire—"this
astoundingly beautiful amber-haired tall slim girl… this luminous creature"
(5)—take turns narrating their time-slipped love. Henry has been flipped into
the past hundreds of times from childhood, stranded

> naked as a jaybird, up to your ankles in ice water in a ditch along an unidentified
> rural route… Sometimes you feel as though you have stood up too quickly even
> if you are lying in bed half asleep. You feel blood rushing in your head, feel
> vertiginous falling sensations. Your hands and feet are tingling and then they
> aren't there at all… and then you are skidding across the forest-green-carpeted
> hallway of a Motel 6 in Athens, Ohio, at 4:16 a.m., Monday, August 6, 1981,
> and you hit your head on someone's door, causing this person, a Ms. Tina
> Schulman from Philadelphia, to open this door and start screaming because
> there's a naked, carpet-burned man passed out at her feet. (viii)

That classic, self-ironic narrative is one of the two voices of the tale, but really both comprise a single civilized middle-class point of view (wealthy upper middle-class, in Clare's case) relaying the kind of love affair seldom recounted outside genre sf, between two entwined lovers and their sometimes wretchedly haywired world. Unlike fantasy treatments of time slippage—by Jack Finney (*Time and Again*, 1970; *From Time to Time*, 1995) or Richard Matheson (*Bid Time Return*, 1975), say—Henry's quirk is eventually attributed by a geneticist to a very rare chromosomal disorder. Chrono-Displacement gets passed on to his and Clare's daughter Alba. The 10 year old child tells him, when they meet, that he has been dead for five years. So Henry knows in advance that he will die when he is 43. While these time travelers have a modicum of free will (or so it seems), they cannot change their timeline to avoid perils, since that would entail altering the past, something forbidden in this novel by the laws of physics. Inevitably, then, the story is deeply tragic—in the mode of fated Greek tragedy but more personal and realistic—and its denouement both disquieting and painfully poignant.

But for a Chrono-Displaced person, the future can sometimes touch the past beyond death:

> She is an old woman; her hair is perfectly white and lies long on her back in a thin stream, over a slight dowager's hump. She wears a sweater the color of coral. The curve of her shoulders, the stiffness of her posture says *here is someone who is very tired*, and I am very tired, myself. I shift my weight from one foot to the other and the floor creaks; the woman turns and sees me and her face is remade into joy… (535)

As is ours, perhaps through tears.

2007 *The Accidental Time Machine* Joe Haldeman

Joe Haldeman (b. 1943) gained a bachelor's degree in physics and astronomy, a solid background for a science fiction writer, and was drafted as a combat engineer in the Vietnam War in 1967. After being badly wounded he was awarded a Purple Heart. Not all of his fiction is based in that experience of war, although his most famous novels were: *The Forever War* (1971) and *Forever Peace* (1998) each won Hugo and Nebula awards. For many years he taught writing at MIT, retiring in 2014, the same year he published his most recent novel, *Work Done for Hire* (where the work is as a near-future hit man). *The Accidental Time Machine* shares some of the grim and confrontational aspects of these works, but this tale of exponential leaps forward in time is above all splendidly entertaining, shot through with wry humor—and an intriguing time machine model.

In 2058 (a date well chosen for the novel's two-generation gap from publication, not too hot and not too cold), deep in icy winter, 27-year old MIT physics doctoral candidate Mathew Fuller presses the RESET button on a new calibrator measuring interactions between photons and gravitons, and the device vanishes from his bench. It reappears almost instantly, too soon for his boss, Professor Marsh, to see it. Naturally, the older scientist assesses this as due to overwork and sleep deprivation. Matt wonders if the diagnosis is correct—what *else* could be responsible—until he hits RESET again and sees the calibrator vanish, then about ten seconds later materialize, shimmer, solidify. Marsh departs, but Matt is determined. Next time, it reappears after a little less than three minutes. Using his cheap Seiko, he times the next effect on its stopwatch function. Thirty-four and a half minutes, or 2073 seconds.

Graphing this sequence on semi-log paper, he finds that each vanishment extends about eleven and three quarter times as long as the previous one. That constrains his further experiments; if he continues, the gap will probably grow to 15 years on the ninth RESET, then 177.5 years, and he will be dead long before it gets that far. The increase is dizzying. On the eleventh reset, if anyone is there to press the button, it will go ahead 2094 years. He has to think this through with great care, starting with a careful experiment using a living creature. He buys a baby turtle, and links the metallic frame of the machine to a metal loaf pan containing his watch, Herman the turtle, a jar of water, and pellets of baby Reptile Chow. Matt presses the button and then waits three agonizing days for the next appointed reappearance.

What follows is at the same time a delightful slapstick romp, a view of a variety of possible and often dismaying futures, a slowly deepening love story, and "some reasonably scientific mumbo jumbo to use as a time machine… gravitons and string theory…" (Author's Note, 277). Haldeman adds:

> When I was about halfway through the novel… an article in *New Scientist* pointed me to a paper by Heinrich Päs and Sandip Pakvasa of the University of Hawaii, and Vanderbilt's Thomas Weiler, "Closed Timelike Curves in Asymmetrically Warped Brane Universes,"[2] which indeed uses gravitons and string theory to describe a time machine. My jaw dropped. (278).

Matt's girlfriend Kara has left him and taken up with a younger fellow whose papers he earlier graded and has now been hired by Professor Marsh as

---

[2] https://arxiv.org/abs/gr-qc/0603045 (2006/09).

his replacement. Driven by a mix of gloom and curiosity, Matt decides to try time hopping in person, which will require a Faraday cage to protect him from any stray radiation. His recreational drug dealer, Denny, has just finished a brilliantly red recoating of a 1956 Thunderbird, and Matt borrows it for a brief trip to the future. He wears a wetsuit just in case the device dumps him in the ocean, and a yellow raft. Of course the trip places him, looking insane in his rubber suit, in the middle of Mass Avenue in a snowy morning rush hour, traffic tearing past him, and is instantly arrested for Denny's murder and the theft of his T-bird.

He is released only because an unknown benefactor pays a million dollars for bail, and leaves him a note via courier: *Get in the car and go.* This must be a future version of himself; Matt talks his way into the pound to retrieve his calibrator, hits RESET, and is immediately 15 years later in a sport field surrounded by thousands of people cheering from bleachers. An aging Professor Marsh greets him with news that Matt is now an honorary professor at MIT, as participant in Marsh's Nobel-winning discovery of time travel. He is quickly a fish out of water, and flings himself another 177.5 years forward after first purchasing some items that might interest antiquarians.

Stealing a cab, he resets into a distressingly degenerated community—not the Mad Max kind, but rather a return to pious ignorance. MIT has now become: the Massachusetts Institute of Theosophy, under fundamentalist professors and priests who ensure that students avoid the sinful study of relativity and quantum mechanics. Matt is accepted as a time traveler and allocated winsome, innocent, intelligent and resourceful Martha as his "graduate assistant," a sort of hand servant to the advanced students. The destruction of almost all advanced technology has been required by Jesus. Allegedly, Christ returned to Earth for the One Year War some three generations earlier, and still holds occasional face-to-face interviews with everyone at the new MIT, including the disbelieving Matt who recognizes a hologram fake when it tries to kill him.

With Martha he escapes still further forward, into a universally wealthy, vapid, safe world under the care of advanced artificial intelligences. A holographic AI named La (the tutelary spirit, so to speak, of Los Angeles) outfits them with a spectacular all terrain mechanical insect, a time machine fitted out with ferocious weapons. What follows is a flight to forever, with stops to visit a terraformed Moon, an Earth with humanity departed and replaced by upgraded bears, and the discovery that La intends to drive the time machine to the end of time, just to learn what is there, even if that means marooning or outright murdering Matt and his new love.

How is all this happening? In a fantasy, it would just be magic, the utterance of a spell or perhaps the sacrifice of a virgin. Haldeman is an expert at finding a rationalization for his physics novelties, the "reasonably scientific mumbo jumbo" mentioned earlier—which in this case is not very far from the real theoretical science of Kip Thorne and the other physicists who have been investigating possible paths to time travel. La explains that

> our space-time continuum is a four-dimensional brane floating through a larger ten- or eleven-dimensional universe… there are countless other four-dimensional branes, but what's important are the five-dimensional ones that can be made to envelop ours… Your broken graviton generator attracted one of these beasts and apparently made a permanent connection… to a huge singularity in our brane: the heat death of the universe. (198–199)

It is this monstrous source of energy that has enabled the damaged generator in Matt's bread-loaf sized set-up to accelerate its passage into the future faster and ever faster. How this narrative closes its time loop must be left undisclosed, but we do already know that some kind of traveler intervened to spring Matt from jail and advise him to take the injured Thunderbird (not a DeLorean in this story) into his early future, on the path that will finally return him, with his new love, back to the past.

2008 *In War Times* and 2011 *This Shared Dream* Kathleen Ann Goonan

On December 7, 1941, "a date which will live in infamy," Japanese military made an undeclared act of war on US forces stationed in Pearl Harbor, Hawaii, killing 2335 soldiers and sailors and 68 civilians, sinking the Battleship USS Arizona with the loss of 1104 lives. On the evening before the attack, in Kathleen Goonan's first of two novels of time shifts and jazz, Sam Dance—an "uncoordinated soldier" with poor eyesight—is seduced by an exotic European physicist:

> Dr. Eliani Hadntz was only five foot three, though she had seemed taller in the classroom, and Sam had not suspected that her tightly pulled-back hair was a mass of wild black curls until the evening she sat on the edge of his narrow boardinghouse bed. A streetlamp threw a glow onto her pale breasts… He had no idea why she was here. (15)

For the rest of that novel (winner of the John W. Campbell award) and its sequel—jointly titled The Dance Family—Dr. Hadntz slips in and out of Sam's life, twisting history from the bloody path it has taken in today's chronicles toward a utopian alternative that dances to the bebop arrhythmic cadences of the jazz Sam and his best pal Wink love. It can't be coincidence that Eliani *Hadntz's* name speaks to our yearning for an alternative world where the worst excesses of a bloody twentieth century *hadn't* happened. To young Sam, a brilliant but unschooled engineer, she brings the plans for an unexplained device that manipulates time by combining a "parallel spiral" (time's multiple courses, we guess, and the DNA double helix) with the quantum uncertainties of consciousness. Hadntz reflects that

> if human consciousness was the time-sensitive entity she believed it was, this device could be called a time machine... that affected the physics and consciousness of human behavior... It would enable humans to use the constant expansion of the universe, in much the same way that the previously invisible power of electricity had been harnessed and was now put to all kinds of positive uses... (22)

This beautiful Hungarian Gypsy scientist, who has left her 12 year old daughter in Nazi-occupied Europe in order to work with the Allies on an atomic weapon, has withdrawn from the Manhattan project in hopes that her device can redeem the very nature of humankind. For Dance, though, mired in war work on magnetrons that will power weapons and radar, such hopes seem elusive. After his beloved brother Keenan is killed at Pearl Harbor, entombed in the Arizona, Sam is flung into the war effort, pursued by Major Bette Elegante of the OSS who wants to know more about the Hadntz device that Sam and Wink try to build. Of course he falls in love with her.

He witnesses atrocities and their consequences. Dr. Hadntz's conviction is that human nature can be led away from the brutish herd mentality and impulses that create war. "How can people treat one another this way?" she asks. "What I am thinking about is how to remove or change this propensity... this urge to be like all the others and to follow a leader blindly..." (129). Sam objects: "Turn the world into your breeding pen? Isn't that what Hitler is trying to do?" (130) No, says Hadntz heatedly; her way does not involve murder but rather a cure for our stupidity.

What anchors these airy sf speculations is the density of Sam's experience in the war and after it, some of this conveyed through his diary entries. And these in turn, Goonan has been quick to acknowledge, are drawn directly from her own father's writings and war stories. Thomas Goonan tried for a year to enter the military despite his poor eyesight, finally did so, was chosen for special training and worked on much the same advanced electronics as Sam, with Company C of the 610th Battalion. He was shipped to France in 1945, then to Germany, supplying troops on the Rhine with ordnance and equipment, and after the German surrender he and his friends opened a biergarten behind their billet in Muchanglandbach, just as Sam and Wink do, liberating huge quantities of wine, barrels, glasses.

Back in the States, amateur saxophonist Sam hangs out in the best jazz venues of the time, catching Dizzy Gillespie and Charlie Parker, and Monk at Minton's in Harlem, wishing Wink were there, illuminated and astonished by the way these utterly new sounds let them feel and see the world in new ways. Or was Allen Winklemyer actually killed in Berlin? Time is unraveling. Jazz is a parallel to the intended action of the mysterious device that Sam continues to tinker with, even after a test version had melted down into a puddle of metal. Jazz resonates throughout the novel, informing the rhythms of the prose in key moments:

> His brain became a device tuned and retuned by Bird's notes; he was tossed like a plane in a wild storm across the astonishing sky of the man's mind…. [He watched] the man bring the notes out from where they flocked within him, building pressure until they burst forth as complex fragments united by tone, by instrument, by his fast-moving fingers, a blur on the keys of his alto sax. (204)

When dead Wink reappears, as time bursts forth in complex fragments united by a new kind of historical melody, Goonan merges her father's recollections with an aspirational science fiction vision of a world that could be made differently. The hinge point of change will not surprise anyone who was young in the same era as Kathleen Goonan—she was born in 1952—but even much younger readers will feel a shiver to find Sam in a world where "the news was much different. Robert Kennedy was president, and JFK was still alive, a globetrotting philanderer… Once they got into D.C. there was no sign of the highway construction that had threatened obliteration of whole neighborhoods," where Sam's youngest children "had grown up in a world free of the threat of nuclear war, due to the Munich Disarmament Treaty negotiated by Khrushchev and Kennedy in 1964" (339).

Deeper currents move under the surface. In *This Shared Dream*, Bette vanishes from that revised history in 1961, and Sam disappears some decade later. Eliani Hadntz is surely involved in these absences, and the adult siblings, Jill, Brian and Megan, watch how their somewhat redeemed world is embracing a new toy, a shape-shifting toy that enhances empathy and can even put its young and not so young users in touch with alternative histories. But Jill, at 41 a poli-sci PhD student, is traumatized by the conviction that her changes to the world—she went back to Dallas in 1963 to save Jack Kennedy's life, triggering this kinder world—were responsible for the mysterious deaths of her parents. Are they really dead, though? And who is moving about furtively in the Dances' old Halcyon House where they grew up?

And what's with these toy cosmonauts, "Spacies," that are all the rage with the kids?

> Spacies had been manifested by the Device… Contact with Spacies would modify the genetic predisposition for unthinking rage and violence and render them subject to consideration of appropriate actions. Touching Spacies also stimulated controlled neurogenesis and might result in undreamed of artistic and intellectual abilities in children as well as adults. (*Shared Dream* 85)

What's more, the connectivity device called Q, well in advance of iPhones and iPads not yet on the market in 1991, is vastly more powerful than anything Apple will produce for decades. "Strong Q explored and used the quantum-physics basis of mind and consciousness to its own advantage, as if it had personality, an agenda… Classbooks using Q were embedded with an altruistic baseline able to evaluate the intent of the user. Q could not be used for injurious purposes" (29).

Inevitably, though, people and organizations are messing about with time, some of them eager to renew the Thousand Year Reich misusing portals opened by Hadntz (who somehow resembles Jill's comic heroine Gypsy Myra) in the deeps of horrific 1940s' global war. Yet like the jazz suffusing both novels, *In War Times* and its successor embrace, like Sam Dance, time's released melody, and dance toward a happier future.

2008 *Time Machines Repaired While-U-Wait* K.A. Bedford, and 2012 *Paradox Resolution*

Aloysius "Spider" Webb, dedicated and ethical to a fault, started a promising career with the Western Australian Police Service. He rose appropriately in rank and responsibility until he learned of a cabal of senior officers engaging in all manner of vile crimes. Torn but driven by conscience, Spider blew the whistle under statutes guaranteeing anonymity. Of course his identity was quickly revealed and his fellow cops cursed him as a contemptible rat and a turncoat. Not only reviled he was cast out, to the fury of his beastly, self-centered wife Molly whom he rather mysteriously loves, doing all he can to retrieve his marriage including household maintenance for a house that is no longer his home.

Without credentials other than his dedicated expertise as a cop, he is forced to return to school for three years and then apprenticeship to gain certification as a Time Machine Repairer, a low-caste and ill-paid job he dislikes. Still, Spider proves self-reliant and profitable as an employee of the bully franchise owner Dickhead McMahon. Depressive (and who can blame him?), Spider is constantly on the edge of fury, saved only by his assistant Charlie Stuart and the capable, caring front desk woman, Malaria. K.A. Bedford's humor is typically Australian: deceptively straight-faced, subversive, abrasively witty—although in this case not at Malaria's expense but in one rueful throwaway line about her parents' choice of name for their daughter taken as evidence of the sheer stupidity of his lazy clients and almost everyone else he encounters.

In 2027, a series of further misfortunes afflict Spider, not least being the multiplied appearances and cryptic claims of different future versions of himself, but starting with a broken 12-year-old second-hand time machine that seems about to blow itself into quantum pieces and possibly take the whole universe with it. Careful inspections show no reason why this should be so, until Spider understands that this machine is in superposition with another, crammed inside a separate spacetime bubble. This second, hidden machine contains as cargo a murdered and mutilated woman, and Spider's crime-solving instincts are aroused. Using a borrowed government device called the Bat Cave, yet another contained spacetime, the repairmen safely explode the machine, and the game is afoot.

So far so Keith Laumer-meets-Philip K. Dick (and that is high praise in this context—indeed, the novel was shortlisted for the 2009 Philip Dick award, and won the Australian Aurealis award for best sf novel in 2008) . What makes *Time Machines Repaired* stand out in the time machine sub-genre is its author's ingenuity, and the recomplications of the plotting. Sometimes it does seem that too much yeast and mystery additives have been spooned and stirred into the mix, leaving crucial issues frothing but unexplained. Even so, this novel and its sequel deal centrally with a consequence of the Time Machine hypothesis that has rarely been carefully addressed: once tractable time travel has been invented and becomes a household staple for the wealthy and criminal, can any stability be retained? What steps can the authorities take to control this seething cauldron of causality?

Kenneth Adrian ("K.A.") Bedford (b. 1963) emerged as one of the more promising sf writers of his nation with *Orbital Burn*, his grim first novel, published by the Canadian firm Edge in 2003. It begins:

One morning, not long before the end of the world, a dead woman named Lou sat drinking espresso in Sheb's Old Earth Diner, one of the few places still open in the cheap part of Stalktown.

Time Machines Repair was also from Edge, but his home city's independent Fremantle Press released an Aussie printing in 2009. It is less immediately *noir* but gets right down to the everyday hassles with time machine. The fourth paragraph starts:

The last time Spider and Charlie had been called out to look over a broken machine—yet another Tempo—it turned out that a cat had gotten trapped in the unit's engine compartment and died. The deceased cat's bodily fluids had then leaking into the translation engine and complicated things needlessly. (1)

The Tempo is a brand marketed by the Tempus Corporation with head-quarters in Nairobi. "Market experts were always predicting the end of the 'time-travel bubble,' but so far demand remained high." This notion of time machines becoming as generally available and fashionable as an iPhone or toy drone, with repair crews available to deal with sagging cats in the works the way computer nerds will fix your computer, is at once charming and horribly alarming. In this case, one small side-benefit is that the ads are not lying: technicians take your damaged machine away, get the works on the bench, diagnose the trouble and fix it, and then, even if it's taken a week or longer, whiz back into the past and make delivery only minutes after the pick-up.

But what of the hazards of paradoxes? Bedford looks at the way road injuries and deaths are minimized in most nations by strict laws and regulations that require training and periodic inspections (unlike gun ownership in the USA). As we have seen, one way to deal with the inevitable prospect of paradoxes (assuming they don't short-circuit in advance and shut the time loop down) is through universe-splitting, which creates allohistories where the agent of time change is isolated without a history into a very similar universe where his or her presence is not disruptive of itself. For example, your present day might be threatened with doom because of mistakes made in the past, but by going back to a point before these errors have happened you could correct them *in a new, separate universe,* using knowledge hard-won in your original world. The information is not sourceless, as it was in Heinlein's "By His Bootstraps" even if the world in which its painstaking discovery was made is now entirely dead, leaving only you, like Ishmael and Job's servants, alive to tell us.

And yet, perhaps inevitably, even this perception is put in question:

> If he could offer them one piece of advice, Spider thought, it would be to leave the bloody time machines alone. Who the hell thought it a good idea to let time machines become a mass-market item? Whose brainwave had that been? Maybe he should try to find out, go back in time and find that person—and kill him, or her, or them, and take care of the whole problem. Except, of course, that might not do it. The email from the future, the one that contained the attachment detailing just how to build working time machines, had landed in the inboxes of a great many engineers… all at once… Someone way up in the future felt it crucially important to start the proliferation of time travel sooner than it would have done otherwise. *That was the guy to kill…* (307–308)

The elaborate layers and foldings of time and status in this remarkable pot-pourri of time should not be crushed into a brief discussion of this kind, and in any case it would be impossible. People perish and return to life in a world identical to the one they left except for their own spacetime vector. Spider is approached by an old, toughened warrior he thinks of as Soldier Spider, and another he calls Near Future Spider. Against their several bills of advice he struggles, determined not to accept interpretations of events that have yet to happen, from his point of view. He passes into the very End of Reality, where a lone starship stands against a million foes represented by the odious Dickhead. Terrifying beings, perhaps as mindless as virulent viruses, the Vores, Angels of Destruction according to Dickhead, are devouring the very crust of the universe from a higher dimensional perspective.

Meanwhile, Molly is locked into a condition of stasis close to permanent death. It is spacetime opera of Grand Guignol bloodiness, laced with rage and sardonic laughter. The sequel, *Paradox Resolution* (initial titled **Paradox Resolution No Extra Charge**, no less whimsically than the first) carries the saga forward, with Spider finding inside his refrigerator the severed head of the thuggish Deadhead McMahon. Further volumes are hinted at. At the very least, these are imaginative romps through time travel territory, but more than that for the keen student of the Time Machine Hypothesis.

12

Looping Time

2014 *The First Fifteen Lives of Harry August* Claire North

The time machine in the twenty-first century is often an individual exploration, psyche-powered, as in a number of fine and increasingly hefty recent
novels, not all of which we have room to discuss here. In this corner of the
narrative Multiverse are versions of the Replay trope (a term borrowed from
the late Ken Grimwood's reiterating fantasy novel *Replay*, from 1998[1]) where
the time travel is sometimes presented without a manufactured machine or
major project to do the heavy lifting. They include Audrey Niffenegger's *The
Time Traveler's Wife* (2003), Kate Atkinson's *Life After Life* (2013) and its
quasi-sequel, tangled, mixed-up time in *The Lost Time Accidents* by John Wray
(2016), scriptwriter Elan Mastai's *All Our Wrong Todays* (2017), and this 400-
pager by Claire North, a pseudonym of Catherine Webb (b.1986), from
2014. It deservedly won the John W. Campbell award for best sf novel
of the year.

Living among us, unknown to almost everyone who is fated to vanish
sooner or later, are the ouroborans. Named for the snake that swallows its own
tail or kalachakras (a Buddhist Tantric term for "wheel of life") they die again
and again only to be reborn as themselves and with many memories of past
cycles available, in principle at least. "The *mind* is what takes the journey

---

[1] Is it fantasy? I remain only half persuaded, but Grimwood's delicious novel is examined in David
Pringle's useful *Modern Fantasy: The Hundred Best Novels* (1988), which declares that "Despite a superfi-
cial similarity to such recent hit movies as *Back to the Future* and *Peggy Sue Got Married*, no one has
produced a time fantasy quite like this one before" (261). It won the 1988 World Fantasy Award.

© Springer Nature Switzerland AG 2019
D. Broderick, *The Time Machine Hypothesis*, Science and Fiction,
https://doi.org/10.1007/978-3-030-16178-1_12

through time while the flesh decays," Harry August reflects; "we are no more and no less than minds, and it is human for the mind to be imperfect and to forget" (41). Harry is one of them, in 1919 born a disowned bastard child of a wealthy Briton and raised by that worthy's gardener. Literally born again with full recollection of his earlier life he goes insane and kills himself at the age of seven, but returns again and again, growing in knowledge, wisdom, patience and ennui.

He discovers the Cronus Club, whose worldwide members share his temporal quirk—without, in the main, his prodigious memory, which makes him a "mnemonic" (113)—and protect each other, making use of future knowledge to enrich and guard themselves while never daring to change the course of history. As Harry lies dying at the end of one such life, a little girl ouroboran brings him a message from a distant tomorrow beyond his lifespan: the world is coming to an end, and this doom is accelerating in its surge backward in time.

There is no time machine in this impressively imagined and masterful naturalistic novel, other than something genomic and quantum-mechanically awry in the constitution of the kalachakras. In this sense it resembles the cyclic death-to-life stacked histories of Kate Atkinson's *Life after Life*. In that novel, though, the cycle is *sui generis*, driven by sorrows and guilts experienced by just one child-girl-adult-child. Part of what makes North's recursive narrative suitable as an extreme instance of sf time travel fiction is the care it takes to provide rational answers to unknown problems. If you maximally live only through the same years, more or less, can you ever learn anything valid from the past before you were conceived or the future following your death?

Yes, suggests the mind of a science fiction writer. If you die in 1989 at the age of 70, you can with some effort have messages conveyed back to you in your failing hours or weeks by a young child who will not die until, say, 2080 and can memorize a message for you from any post-infancy decade prior to that date. Indeed, further iterations are in principle feasible, although the risk of cumulative errors or unintelligible futuristic novelties might make such attempts unreliable and misleading. Similar chaining can bring you information from the years or decades before your birth. All such maneuvers depend on the existence of mutual aid sodalities such as the Cronus Club, and the care all members take to ensure the survival of their colleagues as well as themselves. Oh, and you are forbidden to break history. Of course Harry and his snake-swallower mirror image adversary, Vincent Rankis, do exactly that.

Somewhat younger than Harry, and significantly more intelligent and driven, Vincent pursues a multi-life scheme to advance crucial technologies that will allow him to create the ultimate reductionist scientific tool: a quantum mirror. Quite how this marvelous Faustian engine works is never spelled out, but it draws on the efforts of many skilled technicians and immense quantities of money. It is, of course, powered by a nuclear reactor. Harry is drawn into this Promethean plan by his love/hate relationship with Vincent, readying himself for a final undoing of this attempted leap into the posthuman condition, and his need to obliterate Vincent from the looping time stream. This can be done only by learning the victim's exact date of birth and then, earlier, just after conception, causing an abortion. This is, obviously, a version of the Grandparent Paradox that bedevils time travel, but at least has the virtue of not requiring the mother to be slain as well.

What is a quantum mirror? Vincent describes it as "a theoretical device for the extrapolation of matter… an entirely different way of comprehending the very building blocks of reality, from which comprehension the whole universe may unfold" (207, 209). Harry sarcastically rejects the possibility (but later changes his mind); such a quantum mirror would

> create a theory of everything capable of answering every question starting with how and finishing up with the far more difficult why, this… miraculous device, is nothing more or less than a do-it-yourself deity. You want to build yourself a machine for omnipotence, Vincent? You want to make yourself God? (209)

Vincent, hungry for a final, effective theory of everything, is dismissive: "Maybe all God ever was was a quantum mirror."

Is such an ultimate machine also a proto-time machine, a kind of genome-tweaking machine able to edit *Homo sapiens* zygotes into time-looping *Homo superior*? Something like this implausible genetic modification is what gave Niffenegger's Time Traveler's Wife's husband and daughter their susceptibility to leaps back and forth through time. In the context of the moral perplexities that wind through North's impressive novel, this question is about as relevant and urgent as asking "How many children had Lady Macbeth?"—which is to say, even in a study like this one of the forms of time travel science fiction, not at all. It is an operating conceit that permits us to wonder how one might respond each time one returned to life with the accumulating wisdom or bleak sense of pointlessness from this schooling. North, or perhaps her publisher, was left pondering this, and the book closed outside its narrative frame with a page of heavy black type:

> If you could go back and give yourself one piece of advice from a life already lived, what would it be?

And we readers are invited (or were, in 2014) to submit a reply, at WWW.HARRYAUGUST.NET

That link is now inactive. Such are the sly irritants of cycling back and forth in time.

2017 *All Our Wrong Todays* Elan Mastai

In this merry Vonnegutian cataclysm,[2] Canadian screenwriter Elan Mastai (b. 1974) takes us from a utopian 2016 on the first reverse trip, to 1965, whereupon very bad things happen to all of us in the diverted history we now share and his own science fictional world has been obliterated. Tom Barren (born on October 2, 1983), who writes this memoir, has mastered learned incompetence via the disdain of his fabulously brilliant father, a classic instance of a toxic Narcissistic Personality. Victor Barren builds the inaugural time machine and plans to send it back to the moment when Lionel Goettreider activates a radically new free power source leading directly to utopia.

Or should, until Tom messes everything up, typically, and clumsily destroys the experimental Goettreider Engine, switching fifty redemptive years of benign history for our less than delightful world trajectory since the midsixties. Now known as John, an architect, he finds his dead mother alive, his father seriously improved as a parent, and that he has a younger, snarky sister. Plus his own architectural firm where he wins applause by plagiarizing the urban designs recalled from his vanished comic-strip reality. (This wouldn't work, because such majestic pylons depended on scientific and engineering discoveries not yet made in our world, but then Vonnegut never worried about such details either.)

But John, who is the heartless alternative outcome of being raised in our world, is not quite buried by the arrival of Tom's mind in John's body. Tom comments:

I didn't *feel* anything. I wasn't shoved to the back of his consciousness, scratching and clawing to come out. I was just *annihilated*. I didn't know I was even gone until I woke up and it was the wrong today. (240)

---

[2] Paul Di Filippo catches this with vivid accuracy: "Its short chapters, its punchy, demotic, self-denigrating prose, its tragicomic ambiance resulting in genuine catharsis and epiphany, and its general fascination with the ways humans can screw up—all these are pure Vonnegut." (*Locus Online*, https://locusmag.com/2017/02/paul-di-filippo-reviews-elan-mastai/).

Shock by shock, Tom is pushed at intervals into this nothingness, and John manifests most disturbingly in cold if not brutal sex with Tom's newly beloved Penny and an offended young office worker. Yet even this is not the end of a Doctor Jekyll/Mr. Hyde seesaw of personalities, because from another history where the planet is almost erased by Tom's error with the free energy machine comes Victor Junior, like an evil angel whispering in the tempted ear of John.

Penny is a delightful creation, a not-quite-exact doppelganger of the first world's obsessionally focused Penelope Weschler who washed out of her aspired role as astronaut, and then again as the would-be first chrononaut because she is just barely pregnant after an uncharacteristic fling with Tom. In her rage she kills herself, taking the one-cell conceptus with her. Her our-world double Penny, after the near-rape by John, is reconciled in due course with Tom and they conceive a son, Tom Junior, for whom this memoir is written.

That is the very barest hint at the complexities and delights or terrors of this delicious time travel mini-saga, in which worlds are ruined, even destroyed, or rechanneled, geniuses are unhinged but driven to impossible success, and Mastai Virgils Tom through a divine comedy of blunders and genuine insights, told with humane wit as well as generic insights that might make an informed reader wonder why this marvelous novel was not a Hugo or Nebula winner nor even a nominee. (Mastai could console himself, perhaps, with the $1.3 million book advance and the promise of a movie adaptation).

For all its science fictional prowess, the novel was received better by mainstream readers who presumably skimmed over the ingenious accounts of how the Goettreider Engine (or rather Goettreider Generator, as its inventor complains) gains its free energy from the rotation of the planet, and why most fictional accounts of time travel are nonsensical. Few writers take account of the sidereal motions of the spinning world, its rotation around the Sun, that star's curve through the galaxy, the Milky Way's headlong cavorting through the dark intergalactic wastelands... If a time machine detaches itself from the local gravitational frame of spacetime in order to change temporal position, it loses its lock on the Earth and all the other centers of force. Gregory Benford was conscious of this problem in *Timescape*, so his cast of scientists dealt with the problem by aiming their tachyonic messages of impending doom at the position in space where the Earth had been at a precise time in the past. Most other fictional chrononauts ignore the problem of being almost inevitably Lost in Time.

Goettreider's device emits *tau* radiation, an unknown force until his first machine goes active and quickly kills him and sixteen witnesses. But the first moment when this occurs becomes burned into the history of spacetime, making a detectable target guiding future travelers back to that inaugural moment. In addition, Tom becomes an *atemporal anchor* in a world subject to *temporal drag*. "The fact of my existence warped chronology to ensure my existence. Whatever the trajectory of events might have been without me, subsequent events aligned to produce me here, the same as I was there" (153). Perhaps this is the same implicit effect that not only allowed but forced Michael Moorcock's fake Jesus to replace the imbecile born of the non-virgin Mary in *Behold the Man*. It is, of course, handwaving of the most blatant kind—but it does what good science fiction must do; it fills in an explanatory gap, as Asimov's Hyperspace did for faster-than-light travel, and his positronic brains and Three Laws of Robotics did to make the science fictional world safe from artificial intelligence.

Not that Lionel's version of time travel, a different design from Victor's, makes anyone safer: "time travel is bad for the human brain. Your body can handle it, more or less, but your mind struggles with the cognitive dissonance" (295). Worse, to return again and again to the same spacetime intersect is to rewrite previous memories, and eventually the "conflicting timelines" corrode "the structural integrity of… neural barriers" and the brain begins "to secrete neuritic plaques to scab over what it interpreted as damage to the gray matter" (296) and finally "curdled into a coarse poison—malignant cancer cells buried deep in the memory centers" (297). Perhaps one reason why we don't find many time travelers visiting us.

Mastai presents an argument that the technophilia of much science fiction (the exponential advances toward true AI computing, the virtual reality distractions from reality, the personal jet packs and flying cars, the yearning for time travel) is a false goal, a noxious ideology. If so, he concludes, "we need new futures" (367), not the recycled and chrome-plated futures sold by pulp sf magazines in the 1930s and updated as the latest opioid for a damaged and degenerating imagination. What kind of new futures? Not the Vonnegut variety, we might hope, in his bleak *Galapagos*, where humankind reverted to pre-*sapiens* stupidity, but at least survived without world-killing weapons of mass destruction. Perhaps Elan Mastai can specify a solution better than that in novels still unwritten. I hope he can find a way, though, to retain the time machines.

2018 *Hazards of Time Travel* Joyce Carol Oates

Noted for the variety and prolificacy of her novels, short stories and other imaginative works, many of them marked by Gothic and horror elements, Joyce Carol Oates (b. 1938) had not published a science fiction novel until her 79th year. It deserves listing in this study in rather the same way *Gulliver's Travels* can be regarded as a kind of sf creation, along with *Through the Looking Glass*—which is to say, by generic appropriation. Adriane Strohl is a 17 year old girl who cannot restrain her curiosity and candid questioning at school in the Trumpian kakocratic dystopia[3] of 2039 (in the Reconstituted North American States or RNAS, which includes Mexico and Canada):

> Presidents of the Reconstituted North American States were heads of the Patriot Party. The general population knew little about them though they were believed to be multi-billionaires, or the associates of multi-billionaires… you were conditioned to "like" them by their friendly, smiling facial expressions and by ingeniously addictive musical jingles that accompanied them, as you were urged to "dislike" other figures. To attempt to learn facts about them was in violation of Homeland Security Information and could be considered treasonous. (171)

Adriane is arrested as a "potential subversive," cruelly interrogated, subjected to brain implantation of a control chip, "teletransported" (i.e. teleported) from NAS-13 back to 1959 for at least four years. There, under the bland name Mary Ellen Enright, her true identity forgotten, she is settled into a hideously pert mid-century school run on Skinnerian lines.[4]

What befalls luckless, confused, unattractive Adriane ("She looked like a convalescent. Some wasting disease… that left her skin… sort of grainy," 51) for the next year or so perhaps reflects the real history Oates might have suffered through in the mid-1950s, as a student in Williamsville South High School in Buffalo, New York—although she worked on the school paper and was regarded as a prodigy, winning a Scholastic Art & Writing Award. Still, by contrast with the total surveillance state of NAS, becoming an Exiled Individual or EI with a chip in your head at the end of the nineteen fifties might be preferable to the totalitarian regime where *potential* trouble-makers

---

[3] "As democratic means the rule of the people, so *kakocratic* means the rule of the bad." *Congressional Record*: https://books.google.com/books?id=oA3oFKigTnEC

[4] B.F. Skinner was a major force in the transitional post-behaviorist psychology in the 1950s, espousing "operant conditioning" based on reward rather than classic pain- or deprivation-driven conditioning. It was effectively demolished following Noam Chomsky's savage review in *Language*, 35, No. 1 (1959), 26–58, of Skinner's book *Verbal Behavior*.

can be vaporized by government attack drones. (Is that kind of utter degeneration from the historic values of the US Constitution believable for a mere
20 years hence? Perhaps not, but this is the kind of time compression sf has
traditionally allowed, whether it was written by Robert Heinlein, Marge
Piercy, Philip K. Dick, Margaret Atwood or George Orwell.)

"Mary Ellen Enright" wanders as a stranger in an appallingly strange land
among the other female Acrady [sic] Cottage residents in Wainscotia State
University, Wisconsin, as they struggle into girdles, smoke incessantly like
their professors, giggle over boys, paint their lips bright red, think and say
nothing new or contentious even without the threat of aerial volatilization. A
group of brave student protestors gathers to denounce nuclear weapons testing, and Mary Ellen feels drawn to their passion and righteousness, but knows
that their treatment by campus cops and enraged jocks and towns folk is
finally pointless. No Soviet nuclear attack will befall the continent in the next
80 years—but she cannot explain this not-yet-historic fact without seem-
ing insane.

A decade older that she is, psychology teacher Ira Wolfman attracts her and
seems to reciprocate. Could he, too, be an EI? This proves to be the case; she
is madly in love with him, and eventually his own drive to escape the narrow,
mindless constraints of Wainscotia leads to his plan for them both to escape
to the west coast. (Oates's wordplay was anticipated by sf critic John Clute, for
whom a "wainscot society" is one where people "live in the margins of a dominant civilization or the literal wainscots of its buildings")[5] Although they have
been warned to remain within a restricted ten mile radius or risk instant repri-
sal, Wolfman claims that this is surely a simple mind-control trick, more
Pavlov than Skinner; how can a police state many years in the future track and
punish them in the pre-internet mid-twentieth century?

They struggle through snow for miles, rats in a maze, and find themselves
back where they started. Infuriated, Wolfman rushes again along the misleading track and a black drone falls from the sky, blasts him with a laser beam as
dreadful as a lightning bolt, and his body evaporates. Mary Ellen is seriously
harmed, and we speculate that her neurochip has destroyed most of her new
hard-won learning. With guidance from his dog, a burly young arts teacher
and sculptor she met at the demonstration finds her near death in the woods
and gets her to medical help; she moves in with him as part of an early com-
mune (she is now "fairly attractive," 321), marries him, the end.

Oh, meanwhile she has been contacted by her uncle Toby, now Doctor
Cosgrove, another EI from the future, who apparently has no restricted radius

---

[5] http://sf-encyclopedia.com/entry/wainscot_societies

difficulty driving from his home in St. Cloud, "just over the mountain" in Minnesota (303). Is the tomorrow-land surveillance extensive enough to discriminate between Toby, Adriane and Wolfman, keeping track of their subservience or otherwise? This is the sort of elementary backstory detail that sf, drawing on the century-enriched toolkit of the megatext, would handle with graceful ease. But Oates and other newcomers to this mode of invention are laboriously reinventing the basics, or failing to do so.

Before his abrupt departure from the narrative, Ira Wolfman finally reaches for a rather obvious alternative to the account of their incarceration offered by the 2039 authorities. What if this whole thing is a hoax, like the psychodrama he has planned—stolen from Philip Zimbardo's famous 1971 Stanford prison experiment[6]—for his next project, a version of concoctions in *The Magi* or *The Matrix* (neither of which he cites):

> You believe that crap? 'Exile'? 'Teletransportation'? 'Zone Nine'? None of this is real, Adriane. It's a construct…
>
> You believed it! All Exiles do!… As a CDS researcher I did VVS work—'Visceral Virtual Settings'—my most ingenious project has been a replica of a small town in Wisconsin, 1959 to 1960… I think it's pretty damned impressive—built absolutely to scale, in space and in time…. Time travel is a preposterous notion, my girl… There is no *past*. There is no *there*… I've come to feel that I've created you. (260–261)

We might suppose that Wolfman's disappearance in a bolt of murderous laser light put paid to that boastful claim. Or is temporal teleportation really nothing more than a fairy tale that falls apart the moment simple logic is applied to the notion? If so, Wolfman has been removed from the Matrix diorama, and Mary Ellen and her soon-to-be-husband Jamie Stiles remain in thrall to his mind manipulations, "in a comatose/hypnotized state. They're feeding you through a tube, and emptying you out through a catheter…" (262).

This reading teeters between old time pulp melodrama and postmodernist evacuation of all certainties, leaving us with Joyce Carol Oates as the palpable scenarist. That would be a disappointing discovery to make, so long after the sportive theoretics of deconstruction that arrived decades ago, claiming this is the universal condition of narrative art. Fiction is made up, who could have guessed? Make your choice! Step into the blue Tardis time machine, or swallow the red pill of disillusion!

Or it's just possible that I am misleading you.

---

[6] https://zimbardo.socialpsychology.org/

2019 *Rewrite: Loops in the Timescape* Gregory Benford

In January 14, 2000, Charlie Moment is 48, historian at dull GWU in DC, twice divorced, paunchy and jaded, with two marriages behind him, and just now dead in a traffic accident. He is not dead, though. It is his birthday, in 1968, he is smoothly muscled with a flat belly, and has just turned 16. His life is being rewritten as it is replayed. What is singular (so far) in this suddenly more fashionable sub-genre is that Gregory Benford's novel is also a replay in another sense, with variations, on his important *Timescape*.[7]

This earlier best selling sf novel (plainly not fantasy) is from nearly 40 years ago and discussed above, but Benford draws this time on quantum string theoretic accounts of spacetime and opportunities for re-entering and revising the past. His return to the multi-reality timescape, with upgrades, elaborates provocatively on the nature of time travel in a Many Worlds' mega-universe.

Rewrite is so recent (as I write) and so complicated in its temporal loops and startling interactions that it would be improper to detail the twists and turns of the plot—the logic of which, I suspect, sometimes has eluded Benford himself. The basic device is familiar from Grimwood's *Replay*, as Benford acknowledges with a polite nod from Charlie. Everyone cycles from conception to death without, in almost all cases, knowing this is our fate. That does not make one's actions and reactions predestined, except in the way our choices are framed by a blend of genomic tropisms and culturally instilled biases and capacities. Still, the vast weight of what Asimov called "psychohistory" makes it highly likely that most volitional acts will resemble their counterparts in the past and future, while natural events such as eclipses, floods, meteoric catastrophes and the like will presumably maintain their die-stamped inevitability.

But as Charlie discovers, becoming aware of this cycle enables one to act differently from the "linears." Recalling all that one has learned in a prior maturity reshapes the index of understanding and effectiveness available to a teenager. With the physical vigor of a healthy teen, he retains the perspective of a man nearing fifty. He experiences deeper, more mature feelings for his parents, his father whose endless smoking will kill him from lung cancer at sixty-six. His mother, sinking in depression, lapses into Alzheimer's at seventy

---

[7] It might be argued that we should add *A God in Ruins* (2015), Kate Atkinson's "companion piece" to her splendid novel *Life After Life* (2013) and even the many sequels to Diana Gabaldon's *Outlander*, but the former two are mystical in tenor and the latter a kind of indefinitely extended romance enabled by a Druidic portal mystery.

and kills herself. He has already made love to his 17 year old girl friend Trudy, and does so again with the technical sophistication he has developed, in a previous future, but he cannot follow through an implicit betrothal because she is bourgeois to the bone, insisting that he study law.

Charlie avoided that trap last time, but now he astonishes his teachers, acing all subjects but especially dazzling a gay English teacher with a short story brilliant by student standards. This worthy introduces him to a radical group of nitwits led by... oh, Elsbeth, some years older than he, who last time around married and divorced him. Things get complicated. With the help of a Hollywood hack who shows him the ropes as a writer, he heads to the west coast from Chicago and quickly develops into a notable scriptwriter because he remembers the best movies of coming years and carefully plagiarizes them.

Meanwhile, he has followed hints to The Society, in New York, where he meets another cycling immortal, Giacomo Girolamo Casanova, the self-styled Chevalier de Seingalt, and Casanova's friend Albert. Einstein has reincarnated sideways, laterally, into a body/brain that is insufficiently competent to grasp the original's genius understanding of advanced physics (he works as a taxi driver) but is impelled to try patiently to carry relativity to its string theory culmination in a physics that unites gravity and quantum theory, quantum entanglement and nonlocality.

Also meanwhile, unknown opponents try to murder him, he is haunted by dreams of a Mexican thug, the Hombre, falls for a sexy woman who looks nothing like Elsbeth while resembling her intensely, and so on to the heights of success in Hollywood, drug abuse, recovery—and explorations of the timescape.

Here is the genuine science fiction core of this baroque adventure. Charlie plans to film the novel *Timescape*, now written by Gregory Benford's identical twin Jim, who answers critics who might find these "data dumps" inappropriate for a novel (which, we are always admonished, should show and not tell). Charlie asks the other Benford:

> How could we explain [tachyon communication] in the screenplay? What sort of physical principle might justify the plot device...? I think the movie audience would find the machines and the tachyons a turnoff. Too complicated.
> So you want to film my novel without the hard-SF trappings?
> Sort of.
> Benford bristles. "Then what would be left?" (180)

Fortunately, *Rewrite* does not stint on explanatory excursions, although inevitably they are brutally simplified for non-scientist readers. If the mind of a lateral like Albert can be entangled with a different brain, how does that work?

> ...what does the [quantum] wave think? The connecting currents? The wave stores information, just as a mind does, and can link brains in separate universes through the information, the mind. Charlie looks out at the foaming sea muttering at the foot of the town and thinks of waves overlapping, summing up into white, salty combers that burst over the heads of unlucky swimmers. Waves of quantum whatever, rebounding from the hard rocks of wholly different universes, carrying their surging energy and information from one hammered shoreline and then out to another, a vast, teeming ocean—or rather, a timescape—forever in ferment. (182)

Albert is originally from the twenty-second century, where time travel is understood and mechanized:

> The quantum of information—physicists who have no ear for music call it a qubit—is like a particle of meaning. One can regard our universe as a sum of all its qubits. Your mind is many, many qubits. A swarm of particles, in a way. When coherently engaged, it can be teleported to anywhere else in space-time. The group wave functions—though I do not like that term—are overlapped. The most welcoming portal for such is a complex mind very much like yours— if there is one. (248)

What can trigger this entanglement? Trauma, Einstein suggests—and this is how Charlie started his cross-universe saga, crushed to death in a car crash. And what makes minds so special? "Mathematically, they have the density matrices with the greatest cross correlations. Minds hold the highest information density of all things. Through entanglement they can grasp across the abyss between quantum universes" (249).

At last, then, we have here a science fiction novel that offers hints of how the time machine hypothesis might arise naturally inside the current best frameworks of physics and neurophysics. Bear in mind that Benford is not proposing a thoroughgoing hypothesis, let alone a theory, of time travel, although the book does contain several segments from an imaginary formal paper on this topic written by Albert. And in an unusual Afterword, Benford notes:

> I'll leave the logic that led me to my work on the present novel now, for ultimately, whether ideas work in fiction is up to the reader. I do want to reassure

readers that the ideas that come up in the second half of the novel emerge from current speculative thinking.

This seems a good place to leave these brief studies of sf time travel stories and turn finally to the question of whether evidence in the real world actually supports the time machine hypothesis.

Part III

A Thought Experiment Is Not a Theory

13

In Search of Lost Time Machines

Cosmic string spacetime contains matter that has positive energy density and is physically reasonable. However, the warping that produces the closed timelike curves extends all the way out to infinity and back to the infinite past. Thus these spacetimes were created with time travel in them. We have no reason to believe that our own universe was created in such a warped fashion, and we have no reliable evidence of visitors from the future. (I'm discounting the conspiracy theory, that UFOs are from the future, and that the government knows and is covering it up. Their record of coverups is not that good!)

Stephen Hawking, "Chronology Protection," 2002, 91–92

It seems to be true that humanity has no reliable evidence of visitors from the future, or the past for that matter, or from universes orthogonal to our own. Many of the time travel stories we have examined explain this absence variously: you can go, but you can't come back; paradoxes will prevent you leaving in the first place, or destroy your machine and you with it even before you get started; you can only go into diverged, parallel histories and once there you are stuck. Or perhaps you *create* the universe that you end up in, and while you can leave it you cannot return to your origin, only to a different alternative world. Or the Time Patrol will nab you, sequester your machine (or induct you into their ranks) and make certain you don't try to kill Hitler because that never works out well.

Someday, in a variant quest to pierce the veil of time, you might build a device for "reading the past," even a past that is only a few seconds before this

© Springer Nature Switzerland AG 2019
D. Broderick, *The Time Machine Hypothesis*, Science and Fiction,
https://doi.org/10.1007/978-3-030-16178-1_13

one, but your voyeurism can't be detected in the past. Most plausibly, if a time machine is ever built, nobody taking advantage of its portal to the past can ever go further back than the first moment it will be activated. Or maybe that will be feasible, if the portal is the immense spacetime-warping gravitational field of a cosmic string or wormhole deep in space, but we have no convenient way of getting to it, certainly not now nor conceivably ever. Paul Davies has remarked:

> Going to the past would require something like constructing a wormhole or stargate as a shortcut between two points in space, and then adapting that for time travel. That looks like cosmic engineering of a supercivilization. It's hard to foresee that we could do that, and definitely not in the near future—but, having said that, we may eventually understand enough about the physics of wormholes and strong gravitational fields, and it could be that there is an easier way to do it.[1]

Another risk with an intriguing consequence, explored by Gregory Benford and other science fiction writer speculators, is that even if you *can* travel back or forward in the same universe you start from, it will be very difficult to get anywhere useful. There's an awful lot of dark emptiness out there in the rest of the cosmos, and the Earth, Sun and Milky Way galaxy are moving away from your time machine at enormous velocities so you will never be able to catch up. Well, unless your machine carries a fabulously commodious and effective quantum computer to set the course for your chase, and surely no well-designed time machine would lack one.

Looking on the bright side, perhaps your machine can be locked (*somehow*—too soon to guess how) into the metric frame of the home planet, like a force field string holding the balloon safely tethered. But still, calibration might be rocky, so it would make sense to outfit you time machine with a dark energy propulsion system so you can first lift into cislunar space, well beyond our atmosphere, and return to orbit when you are ready to go home or to another temporal destination. That way you can avoid materializing inside a mountain, or merge with a kid riding a bike or horse along the street, or finding your forearm stuck inside a wall, as in Octavia Butler's *Kindred*.

These considerations are just a few of a time traveler's favorite concerns, and each of the stories we have looked at deals with one or more of these reasons why nobody can be sure that time travel is impossible. Those stories were chosen for this book just because they struck me as historically meaningful,

---

[1] Davies cited from an interview by Karl W. Giberson at https://tinyurl.com/ybmhlh27

fairly or even brilliantly well-written, and presenting ideas to tease the imaginative mind. Many, many admirable stories had to be omitted for reasons of space, and some of those are listed for the reader's guidance or nostalgic reflection in the References at the end of this book. Few of them, regrettably, are from women writers, because somehow until fairly recently the time machine trope has apparently not appealed nearly as much to women as to men except as frank fantasy.

Of those novels and stories discussed above in a little detail, it turns out that more than 62.3% deal with time *machines* or conveyances, a little more than 11% are about what we might dub "Chronoportation" (moving by an exercise of will), and the same proportion attempt the theme of a time *viewer* device. 7.55% deal explicitly with the Many Worlds theory of the metaverse, 3.8% with a natural time *portal*, and 1.87% each with suspended animation or some kind of psychic mind travel.

From all these suggestions from science fiction, one might be worth pursuing seriously as time machine candidates that *have* been seen, perhaps for thousands of years, even if they remain outside the accepted purview of authoritative science.

> CAUTION! Do keep in mind, while reading this chapter, that it presents a thought-experiment in the tradition of science fiction, playing with half-baked ideas and dubious evidence *pour le sport*. I am not asserting that all of what follows is *true*. I wear no tin-foil hat!

Intriguingly enough, the speculation Hawking blithely dismissed in his Chronology Protection paper strikes a familiar note to anyone with even a passing knowledge of the still-formidable and in my view unresolved puzzle of unidentified flying objects, UFOs. True, the vast proportion of UFO reports, as even ardent believers agree, are readily explained as misidentifications of known phenomena. However an intractable residue eludes the most stringent attempts at orthodox explanation. Craft-like UFOs (whether traditional saucer-shaped, sizable triangular craft, or luminous shape-changing oddities) reportedly flick into being, and vanish just as abruptly if too much interest is displayed toward them. Some UFO phenomena are visible to a few witnesses but not all, and yet are associated with genuine, verifiable physical effects.

And in a few extreme cases, made famous in fictional form in the Spielberg movie *Close Encounters of the Third Kind*, plus a virtual epidemic of books and videos about the purported "abduction phenomenon," numerous witnesses claim to have been taken on board unidentified craft and often treated with cold scientistic probing or worse. Their allegedly suppressed memories of these encounters are often recovered (or "recovered") only under subsequent depth hypnosis, a frequently misleading procedure that is just as prone to insert rather than extract preposterous experiences.

Has any serious scientist ever taken any notice of this, you might ask? And even if some have, is there really any relevance to our topic, time machines? Well, here's an ambiguous example: At one time the most widely-known of all astrophysicists was the late Sir Fred Hoyle. He promoted a number of apparently crazy ideas, as well as some, like the once-famous Steady State model of the universe, that proved to be simply wrong. On the other hand, his brilliant intuition led him to a major valid prediction in nuclear physics, an unexpected resonance state of carbon-12 that allows life to exist. Hoyle suggested explicitly and without embarrassment that information from the future can and indeed must affect the present and the past (*The Intelligent Universe*, 1983, 211–215). This does not mean that he thought UFOs are time machines; indeed, in the same book he dismissed UFO reports as errors at best (143–460).

Most of those few scientists who have concluded after close study of the data that UFOs are craft (including, astonishingly, the US Air Force's principal astronomical debunker between 1947 and 1969, the late Dr. J. Allen Hynek, 1910–1986) leaned toward the hypothesis that any such craft must be extraterrestrial. But a moment's reasoning suggests that UFOs would be unlikely to hail from deep space. The late Philip J. Klass (1919–2005), once the best informed and scathing of the skeptics, sardonically emphasized the diversity of well-reported UFO designs and shapes. Unless each UFO is a customized model, such diversity seems to imply the fascinated and simultaneous interest of thousands of distinct interstellar civilizations.

This might be regarded as exceedingly improbable, as is the notion that totally separated evolutionary lines should have produced the somewhat human-like occupants described in the majority of close encounter reports. (Of course, advanced alien genomics or robotics would presumably make it easy to *emulate* hominin appearance—but if an alien is going to take that much trouble to look human, why do so many of the reports claim they are small grays with huge dark eyes, or reptile people, and so on?)

Hynek eventually adopted a more mystical sense of the phenomenon, as did Dr. Jacques Vallée, a French-American astrophysicist turned computer expert and venture capitalist. Vallée argued that related sightings and even beings are frequently found in ancient myths, and suggests that the purpose of UFO entities and their imposed spectacles and control systems is to manipulate our cultures with chaotic intrusions into human psychological verities.

If UFOs are from the future, though, these difficulties evaporate. In the endless eons that lie ahead, there may be many human societies which build time machines. No doubt design would undergo modification, reflecting diverse tastes and technologies. A time craft from a million years hence would not closely resemble the first prototype, or the advanced retrofitted DeLorean of only a century on.

Of course, both explanations can be combined—and there's the added factor, often overlooked, that advanced nanotechnology could reshape any craft almost at will, on the fly, as could the kinds of light-wrapping stealth techniques military forces are already developing.

It has often been suggested that our political leaders, or at least their covert intelligence agencies, know for certain that UFOs are craft but refuse to announce their findings because the truth would unhinge us all. Time machines fleeing from an utterly cataclysmic future would fill that bill better than just about anything else I can imagine. (One slightly more appalling suggestion is offered in explicitly fictional form in my novella "The Womb," incorporated into the novel co-written with Rory Barnes, *Dark Gray*. I won't spoil the grisly revelation by giving it away here.) Fortunately, I think horror versions of that kind are a remote possibility.

The hypothesis of extraterrestrial life, as Carl Sagan often pointed out, is just an idea whose time has come. It is now a routine element of madly popular science fiction movies and computer games. One of the billionaire founders of Microsoft, Paul Allen, funded SETI, the search for extraterrestrial intelligence, with his immense, multimillion dollar Allen Telescope Array northeast of San Francisco. One corollary of this sociological truth that the alien is now utterly familiar can be seen in the readiness people show to interpret any mysterious or misunderstood celestial stimuli as controlled craft, rather than simply accepting them as natural wonders no more remarkable than the equally marvelous Sun and Moon.

The UFO topic burst back into the news in December, 2017, when *The New York Times* and other serious sources reported that after many years of denial, the Pentagon finally admitted that it had funded a highly secret project on anomalous aerial phenomena. $22 million were spent on the Advanced Aerospace Threat Identification Program. One of its particularly striking items of evidence was a Navy F/A-18 Super Hornet's cockpit footage, from 2004, of an unknown white flying object, estimated at some forty feet long.[2] It became known as the "Tic-Tac" sighting, because of its shape. It was "surrounded by some kind of glowing aura traveling at high speed and rotating as

---

[2] https://www.nytimes.com/2017/12/16/us/politics/pentagon-program-ufo-harry-reid.html

it moves. The Navy pilots can be heard trying to understand what they are seeing." (*NYT*: report by Ufology notables Helene Cooper, Ralph Blumenthal and Leslie Kean).

What's more, "For two weeks, the operator said, the [Naval vessel] Princeton had been tracking mysterious aircraft. The objects appeared suddenly at 80,000 feet, and then hurtled toward the sea, eventually stopping at 20,000 feet and hovering. Then they either dropped out of radar range or shot straight back up."[3] Such "sightings were not often reported up the military's chain of command, [former US Senate majority leader Harry Reid] said, because service members were afraid they would be laughed at or stigmatized."

For years, the *New York Times* reports revealed,

the program investigated reports of unidentified flying objects, according to Defense Department officials, interviews with program participants and records obtained by The New York Times. It was run by a military intelligence official, Luis Elizondo, on the fifth floor of the Pentagon's C Ring, deep within the building's maze.

The Defense Department has never before acknowledged the existence of the program, which it says it shut down in 2012. But its backers say that, while the Pentagon ended funding for the effort at that time, the program remains in existence. For the past five years, they say, officials with the program have continued to investigate episodes brought to them by service members, while also carrying out their other Defense Department duties.

Arguably, the most compelling non-classified documentation of the reality of such advanced craft is Leslie Kean's *UFOs: Generals, Pilots, and Government Officials Go on the Record*, published in 2010. It provides statements from highly ranked military officers and scientists, such as Belgium's Major General Wilfred De Brouwer, former Deputy Chief of Staff of the Belgian Air Force; French Air Force Major General Denis Letty; NASA-Ames Research Center scientist Dr. Richard F. Haines; and other well-placed aviation experts. The most remarkable item of UFO evidence is discussed at some length by De Brouwer: a ventral photograph of a triangular object with glaring lights at the three vertices plus a bright central light, all against a black background of night sky. Careful photometric analysis reveals the full effect. Slightly overexposing the image brings up a triangular outline, with a faint glow at its edges; when this is highlighted, a "halo" of light is shown beyond the black shape of the craft, "suggesting the presence of a strong magnetic field" (photos between pages 178 and 179). A similar aerial craft was witnessed at length in late 1989

[3] https://www.nytimes.com/2017/12/16/us/politics/unidentified-flying-object-navy.html

by two policemen, and a total of 141 other independent sightings in a small region were reported when Maj. General De Brouwer was Head of Operations of the Belgian Air Staff. Many other instances and analyses are presented in this informative and cautious book.

We might conclude, then, that there actually are strange vehicles operating in the skies, capable of observing and responding to human-built and operated warcraft, and effortlessly outmatching them. Alien spacecraft after all? Not necessarily.

To date, any possible explanatory link between UFOs and time machines has been restricted largely to speculation, on the assumption (supported by the case just discussed) that perhaps some UFO reports do arise from genuine sightings of inexplicable craft. There is some backing, decades old, for the suggestion that UFOs might be time machines, rather than (or as well as) space machines. The case of Corporal Armando Valdés is one of the most graphic but also, as we shall see, somewhat questionable.

Consider the following account from Santiago, Chile, carried by the world press on Tuesday, May 17, 1977, three weeks after Valdés' alleged April 25 abduction and time contortion. I reprint it in full, as I read it in my newspaper:

> A Chilean Army corporal was kidnapped for 15 minutes by a flying saucer, according to a report from the town of Arica.
>
> An army patrol, led by Corporal Armando Valdés, was on routine duty on the Bolivian border when they saw an intensely bright light about 500 meters away.
>
> The corporal went to investigate and disappeared a few minutes before the object vanished.
>
> Fifteen minutes later, the corporal suddenly appeared amid his men and collapsed unconscious.
>
> The report said his chin was covered in several days growth and his expression was that of utter surprise.
>
> His watch had advanced five days.

If the report were demonstrably true, the experience of diminutive (5′3″) Second Corporal Armando Valdés Garrido at around 4:00 a.m. in Pampa Lluscuma near Putre in Chile would stand among the strongest evidence to date for the Time Machine Hypothesis.

For how else could Valdés have compressed days of physiological change into a quarter of an hour? If the events were reported accurately, it would seem unlikely that the "intensely bright light" seen by the patrol was actually, say, the planet Venus. Nor do weather balloons accelerate the growth of a beard, or fiddle with a watch (although, granted, a prankster might easily change the watch setting). For that matter, it is not a phenomenon known to students of psychosomatics.

And if the Valdés UFO were an extraterrestrial starship, the very best it could manage would be to *squeeze* time. Travelling close to the speed of light, a time-dilated inertia-secured starship might return its kidnap victim five days later despite his conviction that only 15 minutes had passed. But Valdés was allegedly subject to the contrary effect. Allegedly, the Chilean was taken aboard a craft, stayed there for five days, and then was fetched back through time to a point only 15 minutes later than his capture.

A more detailed account was provided the same year by APRO (the Aerial Phenomena Research Organization, shuttered in 1988).[4] Valdés and his six man "patrol" sat chatting or singing beside a campfire, two of them keeping lookout, when Private Rosales reported that a pair of bright violet lights with a red dot at each end had landed, one still visible and casting significant light, approaching and retreating. Frightened, Valdés prayed aloud and ordered the thing to depart, then approached the object and disappeared for a quarter of an hour. When he returned, he reportedly said "You don't know who we are or where we come from but we will be back soon," and lost consciousness for two hours. Members of the patrol noticed that despite his clean-shaven appearance prior to the event, he now had several days of beard. His calendar watch showed 4:30 a.m., although it was now 7 a.m., and the date was advanced from the 25th to the 30th.

The APRO report claims that Chilean fascist dictator (then-President and Commander in Chief) Augusto Pinochet forbade further interviews. Medical, psychiatric and possibly hypnotic tests were planned. Valdés reportedly stated:

> The surprising thing was the way it approached us. As soldiers we are trained to deal with any situation. But this phenomenon didn't seem to have any logical explanation. I would like to regain my memory of those fifteen minutes. I would even like to submit to hypnosis to draw out information about what happened.

The rational starting point in any serious evaluation of this claim must seek answers to basic questions: Could any of this be true? Is the evidence trustworthy? Does it hold up more than four decades later? Do Valdés' crew still stand behind his—and their—earlier claims?

In May, 2004, a somewhat more stringent document was released as "Magonia Supplement No. 50"[5] titled "The Downfall of Corporal Valdés" by

---

[4] "The Chilean Abduction" report is available at http://www.ufoevidence.org/Cases/CaseSubarticle. asp?ID=823

[5] Available at http://www.users.waitrose.com/~magonia/ms50.htm

Diego Zúñiga C., editor in chief of the Chilean ufological bulletin La Nave de los Locos ("The Ship of Fools"), translated by Richard W. Heiden. It establishes its standpoint immediately: "multiple incoherencies in the account make the 'abduction' of Cpl. Armando Valdés Garrido a complex and dubious story."

An immediate discrepancy: here the patrol comprises *eight* men, and not seven (including Valdés). The temperature was below freezing, and the patrol took cover in or near a stables, with two members some 30 feet away guarding the entrance to keep the horses from escaping. A light is visible on a hill and two "stars" descend, one hidden by the hill, the other approaching to within some 500 m, "an oval shape, some 25 m (82 feet) in diameter and with a violet color with two luminous points of a deep red color." Valdés warns it off, then moves toward it, disappears in the darkness, cannot be found for some fifteen minutes, reappears, collapses, is carried to the remnants of the fire, awakens to utter the mysterious message. His watch showed the wrong time and date, but was subsequently lost (?!). Valdés had a horse saddled and he rode the half hour trip to report at his regiment.

An amateur Ufologist quickly picked up the scent and interviewed the conscripts, contaminating their testimony by showing them a book containing pictures of alleged unidentified craft. Since Valdés claimed that he had no memory of any communication with beings associated with the craft (if that's what it was), the general acceptance in the 1970s of an extraterrestrial explanation is plainly a cultural artifact of the period, when UFOs were all the rage in Chilean press and television.

Valdés withdrew from demands for interviews, although as time passed he changed his mind about that. By 1993, promoted to sergeant, he told a regional cable program that the aliens would return, "and it will be necessary to be at peace with God," adding that when his daughter was ill he called upon the aliens to cure her and she recovered. He was, it is worth noting, a member of a fundamentalist evangelical religious sect. This religiose coloring of his tale increased in 1999 when he declared that with the new millennium God's plans would be revealed via a "message" which "will be good or bad according to the point of view from which humanity takes it"—not precisely a falsifiable claim.

He proposed to publish a book on his experience and its consequences, but that has not appeared. The narrative grows increasingly tangled; details are traced in the Magonia article. Interviewed in 1999, one member of the patrol reports additional visitations by aliens, including "escorts, tall, with short white hair, blue eyes, and with the lower part of their bodies similar to a kangaroo's." These beings predicted that 2004 would see "something big," possibly a nuclear explosion. Had these prognosticators specified the Fukushima disaster of March, 2011, or even the New York terror attack of September 11, 2001, we might have felt encouraged to fit this tale a little more securely under the Time Machine Hypothesis...

On the other hand, these accretions have no necessary bearing on the original stark report, and the implication that Valdés was propelled five days into his future and then returned 15 minutes later. Unless this was entirely concocted, or the result of some kind of unusual psychosis (shared by the other conscripts) as proposed by a psychiatrist who interviewed Valdés, or an extremely rare natural phenomenon akin to a spacetime wormhole, we are left with a story that hints at time machines without going much further to support or refute it.

A temporal anomaly of another kind—depending on simple chronology—casts an interesting light on one of the most famous "close encounters of the third kind," that of policeman Lonnie Zamora (who died in November 2009 still maintaining the truth of his encounter) outside Socorro, New Mexico, on Friday April 24, 1964. Professor J. Allen Hynek, retained by the US Air Force Project Blue Book as a scientific consultant, conducted an on-the-spot official investigation of this case within a few days. (We now have some official reason to believe that Blue Book was authorized in part to help cover sightings of top secret military aircraft.)[6]

Hynek declared the witness "a policeman whose character and record were unimpeachable. Physical traces were left on the ground; and, as I personally observed, some of the greasewood bushes in the immediate vicinity had been charred. Even Maj. Quintanilla, then head of Blue Book, was convinced that an actual physical craft had been present."[7]

Also quickly on the spot was Ray Stanford, an unofficial investigator with the endorsement of NICAP, then a leading civilian research group. In a study of the events published in 1978, Stanford claims to have located thirteen witnesses to the Socorro landing, and to have found metallic traces of the craft which were analyzed by Dr. Henry Frankel of NASA's Goddard Space Flight Centre (apparently with no very enthralling result).

Zamora's brief sighting was of an egg-shaped, shiny aluminum-white object, with landing gear and insignia. It was not a NASA test vehicle. Hynek declared bluntly: "Maybe there's a simple, natural explanation for the Socorro incident, but having made a complete study of the events, I do not think so.

[6] https://www.huffingtonpost.com/entry/cia-x-files-flying-saucer_us_56a683cee4b0404eb8f28845
[7] Dr. J. Allen Hynek, *The Hynek UFO Report* (1977, Sphere Books), 223.

It is my opinion that a real, physical event occurred on the outskirts of Socorro that afternoon on April 24, 1964."[8]

Not far south-east of Socorro is the White Sands Missile Range, once the proving grounds for early rocketry experiments. Ray Stanford stated that a five-man team equipped with theodolites was tracking a small balloon from White Sands when they spotted a UFO. "It was easy to see that it was elliptical in shape," said a crewman, "and had a whitish-silver color." The UFO was tracked for a full minute before accelerating away at an estimated 25,200 miles per hour. The team, headed by Navy Commander H. B. McLaughlin, witnessed the UFO on the morning of Sunday, April 24.[9]

But the year was 1949.

Is it likely that a spacecraft would return to almost the same geographical location 15 years to the day later? The probability of chance coincidence is obviously one in 365 by 15 years by the likelihood of being in the same location. Customary standards in science put odds of one in a hundred as the cut-off limit for explanations based on chance coincidence. Of course, prankers might have arranged the second sighting to echo the first. The case for a hoax by nearby students at the small New Mexico Institute of Mining and Technology was revived in 2017.[10] Stirling Colgate, former President of the Institute, had informed double Nobel Laureate Linus Pauling that he knew the name of the student who had arranged this elaborate and well-timed hoax to punish Zamora for his persecution of students driving too fast in the locale. The student was already gone and Colgate declined to identify him. Nobody has since come forward to confess or to out the alleged prankster.

But if the two sightings *were* genuine, and of the same craft, flipping in and out of time, what rationale lies behind the precise 15 year gap? Serious investigators have long wondered whether the appearance of UFOs is really as sporadic as a jumbled catalogue of sightings suggests at first glance. Could UFO occupants deliberately but subtly be drawing attention to their status as time travelers? Or do they just not care, which might suggest visiting aliens from the future who have no reservations about messing with human history?

Is there any discernible pattern? Some six decades ago, the veteran French researcher Aimé Michel (1919–1993) noticed an unexpected regularity in the UFO reports generated in a frenzy over his country in 1954. This impressive

---

[8] *The Hynek UFO Report,* 229.

[9] However a Navy veteran told me: "Tracking moving objects, establishing range and speed CANNOT be done with theodolites, and couldn't even be done with a team of five people in one minute, or 10 minutes, unless they were trained and ready to go with the proper instruments, under close supervision when one appeared."

[10] "Lonnie Zamora and the Socorro UFO," by Brian Dunning: https://skeptoid.com/episodes/4582

wave of sightings stimulated official interest in France which continued unabated, as M. Robert Galley, a former Ministre des Armées, admitted publically on February 21, 1974. Michel found that when sightings for a given day were plotted on a map, they fell predominantly on a series of straight lines. Sometimes several lines intersected at one spot, forming a pattern like an extended asterisk. When such an intersection was itself the source of a sighting, it was usually of a disc standing in the sky, or descending like a dead leaf.[11]

On October 7, 1954, for instance, lines criss-crossing France were generated by 28 reports submitted to the Press, and all fell fairly close (as far as could be estimated) to one of 13 lines. The town of Montlevicq, near La Châtre, was the point of intersection of no less than five of these lines. One line was more than 900 km long, stretching diagonally across France from the Channel to the Mediterranean. These alignments were not stable; they altered drastically from day to day. But it seemed that the grids could be matched with some success by rotating the asterisks.

Preferred flight paths for one or more visiting starships conducting a survey? If the alignments are not due to chance, this is the most obvious (if technologically improbable) guess. But it cannot be the correct interpretation, for a single startling reason:

The sightings along a given line were not arrayed in time sequence.

If a line was indeed a flight path, then the UFO craft must have been shifting to earlier and later points in time as it progressed through space.

In short, if the Michel alignments (which have been derived also for North Africa, the United States, Brazil, Argentina and Spain, and perhaps other countries) are not illusory, the UFOs that generate them would have to be capable of time travel.

Computer specialist and UFO expert Dr. Jacques Vallée set out to estimate the likelihood of the Michel alignments arising by chance. If truly random dots were positioned on a map of France, how many lines would be needed to join them? Would asterisk-shaped constellations emerge, like those which had galvanized Michel? Dr. Vallée's results were equivocal (discussed at some length in his and his wife Janine's 1966/1974 study *Challenge to Science*), but tended to deflate Michel's hopes:

For any distribution with more than 25 points, the percentage of isolated sightings became practically nil. All the sightings were then situated on alignments,

---

[11] Discussed, for example, here: http://www.nicap.org/books/coufo/partII/chIX.htm

and the probability of the formation of very complex networks suddenly became extremely high. (76–77)

Evaluation then turns on the accuracy with which the position of each UFO was estimated. How many people, viewing a craft as common as an airplane, can offer reliable estimates of its distance and height, let alone of an unidentified object whose size is unknown?

Vallée's simulations varied that parameter over a reasonable range, and the array of chance-generated alignments closest to Michel's Montlevicq network presumed an accuracy of 2.5 km (78). Since most of the sightings might well have erred by that much, the significance of Michel's straight lines tends to evaporate. However, Vallée left the matter open to further study. His simulation results, he stated, do "not constitute a final invalidation of the existence of some organization such as that hypothesized by Michel. All it does is reveal a conflict between the precision required by the mathematical method and the fragmentary information we have [by the 1960s] on the sightings themselves" (80). Subsequently, though, he appears to have rejected the alignment hypothesis.

Perhaps the only way to be sure would be to ask the UFO occupants themselves (assuming that they exist). Lonnie Zamora caught a fleeting glimpse of two small white-clad figures near the landed Socorro craft, but he had no chance to chat with them. The most suspect of that expanding category "close encounters of the third kind" are alleged contact cases, face-to-face confrontations and even "abductions" complete with rectal probing ("close encounters of the fourth kind"). Without doubt the most elaborate and sometimes startling of these claims have emerged in sessions of deep hypnosis following time lapses in the consciousness of people who recall only "conventional" sightings.

This is a realm of reportage where the effort needed to sustain open-mindedness, and suspension of disbelief, is almost insupportable. Yet the effort is worth making. Numerous collections of contact reports have become available in recent decades, and from them can be derived a profile of striking underlying consistency. Ironically, the most regular feature is the nonsensical gibberish attributed to the aliens.

After an extensive study "from folklore to flying saucers," Dr. Vallée made these pointed remarks: "The entities' reported behavior is as consistently absurd as the appearance of their craft is ludicrous. In numerous instances of verbal communication with them, their assertions have been systematically misleading" (*Passport to Magonia*, 161). Nor can one readily subscribe to the glib retort that this is due to the poverty of imagination of hoaxers. Many other parameters of the reports, issued from individuals in many countries who cannot have cribbed from one another, are surprisingly consistent.

In Kurt Vonnegut's superb farce, *The Sirens of Titan* (1959), a robot messenger from the distant Tralfamadorians, eons ago, faltered in its programmed 18.5 million year mission, and requires a replacement part. Lacking ready means of communication with its alien masters, the machine provoked the building, by humans, of the Pyramids, Stonehenge and other large marvels of antiquity, which together spelled out a visible cosmic S.O.S. With slowly developing shock, we understand that the entire history of humankind, including the King James translation of Scripture, has been a device conveniently arranged for the repair of a paltry alien space probe.

An early version of the "Vonnegut Thesis" was advanced as long ago as 1962 by the French musician and popular composer Paul Misraki. In *Les Extraterrestres*, Misraki's biblical UFO agents are not simply missionaries or repairmen. They are demiurges, the *2001*-style creators of intelligent life on our world. Erich von Däniken subsequently grew rich by propounding a vulgarized version of this theory, though he lacked Misraki's charming virtuosity. For Misraki, the ancient trials of the Hebrew people constituted, by an appalling irony, a Nazi-style breeding program that culminated successfully in the birth and mission of Jesus, a "superman" mutation with psychic powers. (Geneticists tell us that, strictly speaking, it is impossible to breed for a beneficial mutation, which is by definition a random discontinuity in the genetic line; once randomly created, of course, it can then be retained and multiplied by successful offspring.) In any event, the authorities running Misraki's breeding program were the Elohim or demigods rife in most ancient cosmogonic writings. These still watch us today from UFOs, according to Misraki. Their eschatological plans for humankind are spelled out (when suitably interpreted) in the post-gospel books of the New Testament.

The first thing that strikes adult readers of books such as Misraki's is the curious childishness of these supernal entities. UFO conspiracy theories of history entail a series of gambits as simple as childhood games, replete with the savage tantrums, thoughtless brutality and abrupt sentimentality to be found, among civilized humans, in such games. Misraki himself does not reach his intriguing conclusions through any deficiency in maturity and sophistication. They are the ineluctable outcome of that data he takes for his provenance. Nor is that deplorable childishness simply a reflection of early nomadic innocence, for the data include sources much more recent than the Bible and other ancient scriptures.

Indeed, at the heart of Misraki's conjectures are two UFO reports from the early twentieth century that are surely the most widely witnessed and thoroughly documented of that or any age. The second of these was seen for ten minutes by perhaps 70,000 people gathered together for that express purpose, having been advised two months in advance to expect a remarkable event at that day and hour. Nor are religious interpretations tagged belatedly on this sighting by UFO devotees; they were provided by the "contactees" who predicted the event, and which eventually so impressed the hierarchy of the Catholic Church that it authorized a devotional cult around them. More: as a result of this cult, the sighting zone has become a place of pilgrimage and healing, a locus for the activity commonly termed "miraculous."

The principal sighting, which Jacques Vallée considers undeniably a UFO within accepted parameters, is known to believers as the "miracle of Fátima." It was the culmination of several "angelic visions" by three small Portuguese peasant children: Lucia dos Santos, 10, Francisco Martos, 9, and his sister Jacinta, 7. On 13 October, 1917, between 50 and 70 thousand pious peasants, ambivalent priests, reporters and riotous anti-clerical freethinkers came to the Cova da Iria to see if the predicted wonder would eventuate at midday.

Although the nature of the marvel had not been specified, most seemed to hope for miraculous healings of the celebrated Lourdes variety. The day was foul; despite rain, however, at 11.30 local time Lucia told the crowd to put down their umbrellas. The drenched crowd waited for another half an hour. For an account of what happened next, we can call upon the carefully researched synopsis in *Miracles* (1952)[12] by Jean Hellé (pseudonym for journalist Morvan Lebesque, 1911–1970), a source very much less likely than Misraki or Vallée to err in favor of the UFO hypothesis since it bears the imprimatur of Francis Cardinal Spellman, but consistent with recent non-religious Portuguese studies on the events.[13]

At 12.05 pm, Hellé wrote,

the rain stopped abruptly, and in scarcely a minute the last clouds had dispersed. The sun appeared: it was not golden but looked like a silver disk. It seemed to be surrounded by another disk whose color the spectators could not state beyond saying that it was dazzling, though they acknowledged that this effect might have been due for the moment to the sudden appearance of the sun in a sky which until then had been overcast.

---

[12] David McKay Company, Inc., New York.

[13] A far more skeptical trilogy on the topic has been published by Fernando Fernandes and Raul Berenguel: *Heaven Lights, Celestial Secrets* and *Fátima Revisited: The Apparition Phenomenon in Ufology, Psychology, and Science* (2007–2010), a transdisciplinary study by the Multicultural Apparitions Research International Academic Network (Project MARIAN) at the University Fernando Pessoa in Porto, Portugal.

The sun remained motionless for an instant and then, as abruptly as it had appeared, began to tremble. It almost looked as if it was shaking itself. It stopped trembling and then began to spin around, shooting out on all sides rays of light which changed color. These rays were, in turn, red, blue, violet and green; they colored the spectators' faces. The sun stopped spinning, and stood still in the heavens; then it began to spin round, again throwing out colored rays of even greater brilliance.[14] It then stopped spinning for a second time.

A moment passed and then the spectators received the impression that the sun detached itself from the sky. Literally, it seemed to jump in space. It zig-zagged about from east to west and then, as if quite eccentric and, in some sort, crazed, it fell from the sky, plunging toward the earth and giving out unbearable heat. At this moment the feelings of the crowd were divided between wonder and fear. From all sides shouts went up. "Take pity on us! Have mercy on us! It's the end of the world!" The sun came to rest for a third time. Then it zigzagged its way up into the sky and, so to say, back into its usual place. The sky was clear again and with no sign of cloud.

The spectators suddenly realized that their clothes were dry once more.

Needless to say, this erratic celestial object was not the sun. Although it was observed by a schoolmistress and her pupils six and a quarter miles from Fátima, and by the poet Alfonso Lopes Vieria from a distance of 31 miles, the Lisbon meteorological archives record only this: "Storms in the morning, showers; bright periods from 1 p.m. to 6 p.m."

Nor was it a contagious hallucination. Avelino de Almeida, editor of the anti-Catholic leftwing newspaper *O Seculo*, wrote: "From the road above, where the cars were crowded together, and there were several hundred persons standing, the immense crowd was seen to turn towards the sun which appeared, free from cloud, at its zenith. It looked like a dull silvery plaque." There was, in short, no delayed questing and rumor mongering about the alleged marvel: the crowd reacted as one. And, as one, they watched the event end.

Hallé summed up nicely:

The miracle might well have been suspect if for some of the crowd the dance had continued, if each individual had seen something different, or, I may add, if the crowd had made an exhibition of itself. But there was not the slightest sign of nervous disorder, fainting or collective hysteria. Photographs were taken; they show us, in every direction, peasants crowded together...Their faces show merely intense amazement.

---

[14] Note the resemblance to one puzzling element in the "Tic-Tac" UFO filmed by US Navy aircraft, discussed above.

One might wonder why, if "photographs were taken," there are none show-ing the gyrations of the "dull silvery plaque"? Vallée has noted that the rudi-mentary photographic film of that era was not sufficiently fast to capture transient light phenomena.

That athletic UFO was not the first to be seen at Fátima, although it was by far the more spectacular. A month earlier, September 13, 1917, between 25 and 30 thousand onlookers had gathered at the Cova by noon. "The sky dark-ened," states Hellé. "A witness declared, 'It was as if dusk had fallen very quickly and I thought I could make out the stars, as if it were really nightfall.'" According to the testimony of Father Joao Quaresma, a future vicar-general of Leiria,

> to my great astonishment...I saw, clearly and distinctly, a luminous globe com-ing from the east and moving to the west, gliding slowly and majestically through space. With my hand I gestured to Monsignor Gois who was standing next to me, and who had been making fun of me for coming. Looking up he too had the good fortune to see this unexpected vision.

Oblivious to any but the religious import of what he writes, Hellé notes that "It was, in fact, a luminous globe and according to the assertions of those who saw it oval in form with 'the widest part underneath'. 'A sort of celestial airplane,' Canon C. Barthas wrote afterward, in the light of this evidence."

The oval aeroform was followed by "a fall of white petals, which 'like round, shining snowflakes floated down towards the earth, in a strong beam of pre-ternatural light.' Contrary to all the laws of perspective," Hellé writes, "these petals or flakes became smaller as they fell." The white flakes, at least, were photographed by Antonio Rebelo Martins, Portuguese vice-consul at the time to the USA. This sounds very much like a still unexplained phenomenon sometimes associated with flying saucer sightings: "angel's hair," to which Donald Kehoe devotes an entire chapter in a book with the disturbingly appropriate title, *The Flying Saucer Conspiracy* (1955).

The significance of the Fátima UFOs does not lie in their being observed by so many thousands, but in their correlation with "visions" by the children of distinctly theological overtone. If we postulate that these marvels were indeed controlled craft of some sort, their appearance on cue would seem to validate the claims of the visionary child Lucia, later to become a nun. Yet what she reported was the stuff of pre-Vatican II catechisms, of bloody-minded pulpit exhortations about the gruesome fate of the damned.

A series of visions culminating in the October manifestation began on Sunday morning, 13 May, 1917, when the children led their flock to the

Cova. Years later, when she had taken vows in an enclosed order, Lucia stated that this was not, however, the first vision. By the time of this particular revelation both the Marto children were dead, so skepticism may well be justified. Although the children had agreed to tell nobody of the May 1917 vision, Jacinta could not contain herself, and had excitedly informed her mother at the first opportunity. Is it plausible that she would not have done the same thing the previous year; that, in numerous grillings that followed 13 May, it would not have been mentioned?

Lucia's account is that in the spring of 1916, sheltering from drizzle in a cleft of rock, the children saw a young man in white garb who identified himself as "the Angel of Peace," instructing them to pray for non-believers. Two months later he appeared again, asking for prayer and sacrifice, and declaring himself the Guardian Angel of Portugal. A further two months later, he appeared with a chalice surmounted by a host from which blood dripped, gave the children Holy Communion, and vanished, leaving them in a state of lassitude and ecstasy.

The 1917 apparitions in the Cova, attested to by all three children, began with a flash of light in a clear sky. A second flash, says Hellé, "rooted them to the spot. It was not exactly like a flash during a storm, but rather a bright light which surrounded them and which all three imagined to come from the right." Looking in that direction, they saw, "encircled, like them, with a blinding light, a lady more resplendent than the day...

"'Don't be afraid,' they heard, 'I won't hurt you.'"

This dazzlingly bright figure was a girl of perhaps 15, her feet resting on a cloud. She held a rosary in her clasped hands and told the children: "I come from Heaven." After inviting them to return to the same spot on the 13th of each month until October, when she promised to reveal her identity and mission, she informed them that they would all go to Heaven. Nine year old Francisco, though, would "first have to say many rosaries." More was disclosed, some of it not made public for decades. Evidently Francisco did not hear the apparition's voice, although Jacinta and Lucia did. Perhaps this was a function of his "sinfulness." He died two years later from influenza. His early demise, as well as Jacinta's, had been predicted in the vision.

If all this were taken literally, the apparition's disclosures embody a view of moral accountability grim with medieval severity. The vices available to a pious boy fated to die at age 11 are, one might suppose, severely limited—and Francisco was no hellion. Even before the "angelic vision," the children were compulsive rosary-sayers. This drastic calculus of salvation—as blood-curdling as the famous retreat sermons in James Joyce's *Portrait of the Artist as a Young man*—is not the least interesting feature of the Fátima message.

After further prayers, the vision disappeared into the sky, "all at once," "straight up."

It did not take long for the word to get around. On 13 June, 60 or so people followed the children to the Cova, and heard Lucia say, "Look, there's the flash. The Lady's coming." This time, strangely, the girl instructed Lucia to learn to read "so that I can tell you what I want." Lucia proved lax in obeying this order, which in any case would have been difficult for a peasant girl to carry out, and no rationale has ever been provided.

Now the girl stated that the two Marto children would soon die, and asked for "devotion to my Immaculate Heart" to be spread throughout the world. The following month, she again asked Lucia to learn to read, and once more specified that they must return on the 13th of the next month. "You must recite the rosary every day. Say it to obtain the end of the war. Only the intercession of the Blessed Virgin can obtain this grace for men." When Lucia asked for a miracle, so that everyone might believe in the apparition, the girl promised one for October. Certain specific requests for physical healing and spiritual grace were answered with refusal or ambiguity.

Then a quite devastating phenomenon occurred—visible, of course, only to the children, although their terror and exhaustion were manifest to onlookers. The girl opened her hands, projecting beams of light. Lucia did not provide an account of this vision until long after the October UFO "miracle," but it is consistent with the children's reported reactions at the time.

> The rays of light reflected seemed to penetrate into the earth and we saw what seemed to be a great fiery sea in which were immersed demons, all black and burned, and souls in human form like transparent glowing embers. They were raised in the air by the flames and fell down in all directions, like sparks in an immense fire, without weight or equilibrium, with loud cries and groans of pain and despair, making us shudder and tremble with fear.
>
> The demons could be distinguished from the humans by their horrible, repellent shapes as of terrible unknown animals, and were transparent as burning coals.

The girl left the children in no doubt about what they had seen. These were the souls of the damned in hell.

One of the minor oddities of traditional religious psychology is that many Christians who fight tooth and nail to prevent their children being exposed to the cruelty of horror movies find no difficulty in ascribing to the "Mother of God" Lucia's revolting horror show of the eternally damned. Studies of childhood fantasy amply demonstrate that the young are perfectly able to invent their own terrors and monsters, creations that would trouble any adult's sleep.

But the deliberate infliction on children of macabre, repulsive imagery is surely the act of a sadist.

The possible sources of the Fátima "vision of hell," and others like it, are very limited. It can be assigned to one of four categories. It might have been due to the children's collective fantasy, abetted by earlier indoctrination or even coaching by conniving adults. It might be Lucia's morbid adult fantasy, imposed retrospectively. It could have its source in an intelligent though vile entity associated with the well-attested UFOs (perhaps, yes, time machines). Or, of course, it could have come straight from the lips of the Mother of the Christian God.

If either of the first two are true, we must account for the coincidental arrival of the UFOs—no easy task. The latter two possibilities cause the moral sensibilities of our time to recoil in disgust. Either we are trapped in a Gnostic cosmos controlled by an insane god, or, on a less universal scale, in one containing powerful psychopathic UFO or time traveling entities.

By 13 August 1917, appetites were well whetted. Twenty thousand pilgrims assembled; many reported, at midday, a clap of thunder, a flash of light, and a small cloud hovering above the tree for ten minutes. But the children were not there, for they had been detained by the sub-prefect of the province and questioned with threats of boiling in oil (proving, perhaps, that humans are not always more compassionate than UFO child frighteners). Released, they went to the Cove four days later and the illuminated girl made an appearance out of schedule. Peevishly, she told them: "Because of the wicked men and their sectarianism, the miracle promised for October will be less striking." Money left by the pious should be used to buy one silvered and one gilded processional stand; any excess could go toward a modest chapel in her honor.

The September apparition was attended by the first UFO and the fall of "angel's hair." During the final vision, in October, with its "dancing sun" UFO, the girl at last named herself. "I am Our Lady of the Rosary, and I wish a chapel to be built here in my honor." How artless, this queenly dictum; how undisguisedly grandiose.

The content of the Fátima visions—apart from the predictions of a spectacular event—is a blend of the conventionally exhortative, the sentimental, the malicious, the grandiose and the prevaricative. True, the experience had a profound effect on the eldest seer, who spent the rest of her life confined in a convent. On the other hand, this was doubtless a social advance for a peasant girl, and is hardly in their favor in any case: a life of total renunciation is scarcely the human optimum. And, of course, crackpot revelations of a hundred brands have had their saintly adherents. Few others, though, have been preceded by UFO displays or followed by genuine physical miracles. Today,

Fátima, like Lourdes, is a shrine of terminal suffering where, against all reason, cases of inexplicable healing are reported. (Although, oddly enough, no amputated limbs have ever been miraculously regrown…)

Unfortunately, the mother of Jesus, we learn, has been obliged to visit repeatedly to warn us of our planetary and moral danger. Thus, at La Salette (where two children met a "beautiful lady" radiating light, on 19 September, 1846) she stated explicitly: "If my people will not submit, I shall be forced to let go my son's arm; it is so strong and heavy that I can no longer hold it up."

In *Passport to Magonia: From Folklore to Flying Saucers* (1969), Jacques Vallée drew together from history reports of events of this kind associated with flying craft and mysterious messages, warnings and threats:

> We could also imagine that for centuries some superior intelligence has been projecting into our environment (chosen for reasons best known to that intelligence) various artificial objects whose creation is a pure form of art. Perhaps it enjoys our puzzlement, or perhaps it is trying to teach us some new concept. Perhaps it is acting in a purely gratuitous effort, and its creations are as impossible for us to understand as is the Picasso sculpture in Chicago to the birds that perch on it. (160)

Or maybe taunting the ignorant of their past eons is a sport for debauched time travelers—perhaps the same small group from the future whose idea of fun is dropping in on the gullible at regular intervals and frightening them.

With the Time Machine Hypothesis in mind, consider the case of a young married couple, Peter and Frances, whose story was detailed in *Flying Saucer Review*, in March 1975, by investigator Carl Van Vlierden. At the time, Editor Charles Bowen called it perhaps "one of the most important yet to have appeared."

Driving from Salisbury, Rhodesia (now Zimbabwe) to Durban, South Africa on May 31, 1974, the two witnessed a bluish, pulsating aerial light at 2:30 a.m. A sequence of extraordinary events followed, including a period of coma. Arriving at Beit Bridge, Peter consulted his car's instruments and found that the trip meter "had recorded only 17 km, yet the distance by road from Fort Victoria to Beit Bridge is 288 km." What's more, less than two liters of gasoline had been used.

Eventually, hypnotic investigation of the amnesic period was undertaken by a Czech physician, Paul Obertik, M.D., a South African resident. Peter went into very deep trance, and appeared to recall events from the coma period. (Frances remembered nothing extra, perhaps indicating that she had been totally unconscious.)

Two UFO craft, he recalled in trance, had flown above or near their car, taking control of it. An entity "beamed straight to the back seat and sat there the entire journey." Somehow Peter was able to communicate with it.

Where did they come from? asked Dr. Obertik.

"Outer galaxies."

Which outer galaxies?

"They didn't give any names… they just said… outer galaxies."

Later, the doctor asked: "How far can they travel, Peter? How fast? Can they travel faster than the speed of light?"

"They travel by time," Peter said, deep in trance.

"They travel by time? What do you mean exactly by that?"

"They can travel on time… speed of light is too slow to cover billions of miles in seconds. If they want to go from point A to point B they have to come back in time to get to Earth, so they send themselves back into time. They are time travelers, not space travelers."

Still later, Peter stated that the UFO occupants came from twelve planets in our own galaxy, the Milky Way, and do not any longer wage war, being "about 2000 years ahead of Earth."

What's more, they have infiltrated our planet, posing as clerks, businessmen, lecturers and even dustbin cleaners. This was long before the movie *Men in Black* and its sequels, but the idea had a certain similar charm. Interstellar trash men!

"In the light of his considerable medical and experimental experience with hypnosis," Carl Van Vlierden summarized, Dr. Obertik's conclusion "was that Peter experienced the things he talked about." This is a rash judgment for a capable hypnotist to make, since truth or falsity in such trance-evoked memories is always open to question, let alone in such a bizarre case. Still, we may take Peter's testimony more seriously under these conditions than if he had simply trotted out his peculiar tale over a glass of wine.

The details about planets and galaxies are starkly inconsistent, though they reflect the usual origins attributed to UFO occupants in encounter cases ("They come from Zeta Reticuli and the Pleiades" etc). More interesting is the account of why time travel is necessary: to compensate for the colossal distances traversed by the craft. We have seen this postulate invoked by Dr. Asimov in his discussion of wormhole matter transmission, and it is fully in accord with the theory of tachyons, those faster than light particles that, alas, probably don't exist. Unless Peter circa 1974 was up to date on advanced speculations in the field of high-energy physics, it is hard to see how he could have invented this detail.

In 1975, in *The Invisible College*,[15] Vallée commented:

UFO reports are not necessarily caused by visits from space travelers. The phenomenon could be a manifestation of a much more complex technology. If time and space are not as simple in structure as physicists have assumed until now, then the question "where do they come from?" may be meaningless: they could come from a place in *time*. (38)

Interestingly, in 1980, 26 year old Sgt. James Penniston was one of the USAF Security Police team who claim to have seen (and in his case touched) a landed UFO in the now-famous Rendelsham Forest UFO incident, in Suffolk, UK.[16] Penniston was deeply affected by this event, and claims that the authorities initially forced him to sign a very muted and misleading account of his experience. His initial interpretation of the craft and its presumed occupants was that it came from the future, although more recently he suspects that it was "interdimensional," whatever that means.[17]

John Keel (1930–2009) and others have pointed out that around the 1870s, newspapers in a number of countries carried a rash of "mysterious airship" reports. Witnesses sometimes reported chatting to the occupants, who claimed to be airship inventors. Due to an embarrassing scarcity of genuine airships at the time, some UFO theorists see this as a psychological effect: the interpretation of puzzling aircraft data in terms of current expectations (rather than the simpler explanation: wholesale invention by newspaper editors eager to sell the equivalent of Elvis sightings to the rubes).

Keel disagreed. Many reports, he argued, were too detailed to have arisen from simple malobservation. Nor could all of them be attributed to fraud on the part of sales-hungry newspaper proprietors and journalists. For Keel, like Dr. Vallée, the changing nature of UFO accounts is proof of a "Grand Deception." If the UFO occupants (assuming they exist) ever tell the truth, they make certain to stir in plenty of gibberish as well.

If UFOs are craft under intelligent alien control, harking from outer space, it is difficult to see why their occupants would bother with such foolish games. But if UFOs are time machines, it would be psychologically advantageous for their crews to study an era in disguise. What better tactic, if they wish to intervene marginally in our affairs, than to misuse the technological or

---

[15] Pagination cited above from the 1977 British Panther edition reprinted, rather misleadingly, as *UFOs: The Psychic Solution*. A more determined attempt to show that UFOs are time machines is *Identified Flying Objects*, by anthropology professor Michael P. Masters, PhD (2019).

[16] See, e.g., http://www.therendleshamforestincident.com/The_Full_Report.html

[17] He discussed this changing assessment in an interview with Connie Willis on the rather broad-church conspiracy theory AM radio show *Coast to Coast* (January 25, 2019).

religious trappings of the day, thus distracting notice from their true nature and origin?

Similarly, if ordinary time-bound folk do stumble on them (supposing a "paradox loop" escape route does not exist, or fails to occur for some reason), it would be easy to offer some preposterous account that will satisfy the witness while stimulating nothing but derision in official quarters.

By the same frustrating token, it would perhaps be necessary to the long-term success of such a ploy that some witnesses be told that UFOs *are* time machines. A partial admission mixed with confusing nonsense would cloud the trail. The most effective lies always incorporate the incomplete truth.

The real difficulty with this analysis is that it is entirely protected against testing. It is unfalsifiable. There is no polite reply to someone who claims that fairies dance at the bottom of his garden, but that they can't be seen because the presence of witnesses causes them to turn instantly into plaster gnomes. The outright fabrication of testable evidence, such as photographs, used to be detectable, but these days digital CGI wizardry is far harder to unmask, even the home-brewed kind. There is no doubt that "contactee" George Adamski's inane tales (in *Flying Saucers Have Landed* and *Inside the Spaceships*) of handsome long-haired blond Venusians and Saturnians were absolutely fraudulent, as were his bogus but influential photographs of flying saucers and "motherships." In other cases, though, particularly where the alleged witnesses to visitation from space or a Heavenly realm are otherwise credible and reluctant to reveal their experiences, it is possible that their only blunder is to believe what they were told by duplicitous time travelers.

Here, then, is where we stand at the end of our journey:

We have inspected many science fictional varieties of time machines and other methods of time travel, as well as current science as it addresses the possibility that this prospect is more than imaginative sf. A neat summary of the state of play was offered by the late Stephen Hawking not long before his death, in a chapter of his final book *Brief Answers to the Big Questions* (2018), "Is Time Travel Possible" (125–142):

> In conclusion, rapid space travel and travel back in time can't be ruled out according to our present understanding. They would cause great logical problems, so let's hope there's a Chronology Protection Law to prevent people going back and killing their parents. But science-fiction fans need not lose heart. There's help in M-theory. (142)

And even if M-theory fades in turn from scientific popularity, it will surely be replaced by some further theory of spacetime and beyond. Maybe that next theory will contain a yet-unknown supplement large enough to fly a time machine through it.

References, Sources and Further Reading

References to Science Sources, Shown Chronologically

John G. Taylor: *Black Holes: The End of the Universe* (Souvenir Press, 1973)

John Gribbin: *White Holes: Cosmic Gushers in the Universe* (Paladin, 1977)

————: *Time Warps* (Dent, 1979)

————: *In Search of the Edge of Time* (Transworld, 1992)

Fred Hoyle: *The Intelligent Universe* (NY: Holt, Rinehart and Winston, 1983)

Michio Kaku: *Hyperspace: A Scientific Odyssey Through Parallel universes, Time Warps, and the Tenth Dimension* (Oxford UP, 1994)

————: *Visions* (Oxford UP, 1998)

David Deutsch: *The Fabric of Reality* (Alan Lane, 1997)

Paul J. Nahin: *Time Machines: Time Travel in Physics, Metaphysics, and Science Fiction* (NY: American Institute of Physics, 1993); revised and updated as *Time Machine Tales: The Science Fiction Adventures and Philosophical Puzzles of Time Travel* (Springer, 2017)

Kip S. Thorne: *Black Holes and Time Warps: Einstein's Outrageous Legacy* (London: Picador, 1994)

————: *The Science of* Interstellar (NY: W.W. Norton & Co., 2014)

Stephen Hawking: *A Brief History of Time: Updated and expanded tenth anniversary edition.* (NY: Bantam Books, 1988/revised 1998)

Clifford A. Pickover: *Time: A Traveler's Guide* (NY: Oxford University Press, 1998)

J. Richard Gott: *Time Travel In Einstein's Universe: The Physical Possibilities of Travel through Time* (Boston New York: Houghton Mifflin, 2001)

Paul Davies: *How To Build a Time Machine* (London NY: Allen Lane, 2001)

Brian Greene: *The Elegant Universe: Superstrings, Hidden Dimensions, and the Quest for the Ultimate Theory* (NY: Norton, 2003)

Lisa Randall: *Warped Passages: Unraveling the Mysteries of the Universe's Hidden Dimensions* (NY: Ecco Press, 2005)

Ronald Mallet, with Bruce Henderson: *Time Traveler: A Scientist's Personal Mission to Make Time Travel a Reality* (NY: Thunder's Mouth Press, 2006)

David Toomey: *The New Time Travelers: A Journey to the Frontiers of Physics* (NY: Norton, 2007)

Sean Carroll: *From Eternity to Here: The Quest for the Ultimate Theory of Time* (NY: Dutton, 2010)

Dave Goldberg and Jeff Bloomquist: *A User's Guide to the Universe* (Hoboken, 2010) —see Chapter 5, "Time Travel," pp. 131–164

Leslie Kean: *UFOs: Generals, Pilots, and Government Officials Go on the Record* (NY: Harmony Books, 2010)

Brian Clegg: *How To Build a Time Machine: The Real Science of Time Travel* (NY: St. Martin's Press, 2011)

Jim Al-Khalili: *Black Holes, Wormholes and Time Machines* (CRC Press, Boca Raton, Florida, 2012)

Kip Thorne: *The Science of* Interstellar (NY: Norton, 2014)

James Gleick: *Time Travel: A History* (NY: Vintage Random House, 2016)

Adam Becker: *What is Real? The Unfinished Quest for the Meaning of Quantum Physics* (NY: Basic Books, 2018)

Michael P. Masters: *Identified Flying Objects: A Multidisciplinary Scientific Approach to the UFO Phenomenon* (Masters Creative LLC, 2019)

Fiction Works Discussed, Plus Suggested Reading

Martin Amis *Time's Arrow* 1991

Poul Anderson "Flight to Forever" 1950

———— Time Patrol sequence 1955—

———— *The Dancer from Atlantis* 1971

———— *There Will Be Time* 1972

———— *The Avatar* 1978

Isaac Asimov *The End of Eternity* 1955

Isaac Asimov and Robert Silverberg *The Ugly Little Boy* 1958/1992

Kate Atkinson *Life After Life* 2013

Kage Baker *In the Garden of Iden* sequence 1997—

John Barnes Timeline Wars: *Patton's Spaceship, Book 1*; *Washington's Dirigible, Book 2*; *Caesar's Bicycle, Book 3* 1997

Stephen Baxter *Timelike Infinity* 1992

———— *The Time Ships* 1995
———— *Manifold Time* 1999
Barrington J. Bayley *The Fall of Chronopolis* 1974
K. A. Bedford *Time Machines Repaired While-U-Wait* 2008
———— *Paradox Resolution* 2012
Gregory Benford *Timescape* 1980
———— *Rewrite: Loops in the Timescape* 2019
Michael Bishop *No Enemy but Time* 1982
James Blish "Beep"/*The Quincunx of Time* 1953; 1973
John Brunner *Times Without Number* 1962/69
John Brunner and Damien Broderick *Threshold of Eternity* 1959/2016
Damien Broderick *The Dreaming* (revised, Fantastic Books, 2009)
———— *The Judas Mandala* 1982
Octavia Butler *Kindred* 1979
Arthur C. Clarke and Stephen Baxter *The Light of Other Days* 2000
Michael Crichton *Timeline* 1999
John Crowley "Great Work of Time" 1989
Philip K. Dick *Counter-Clock World* 1967
———— *Now Wait for Last Year* 1968
Gordon R. Dickson *Time Storm* 1977
Robert L. Forward *Timemaster* 1992
John Fowles *A Maggot* 1985
Nancy Fulda "Backlash" 2010
Diana Gabaldon *Outlander* series 1991—
David Gerrold *The Man Who Folded Himself* 1973
Joe Haldeman *The Accidental Time Machine* 2007
Robert Heinlein "By His Bootstraps" 1941
———— *The Door Into Summer* 1956
———— "All You Zombies" 1959
———— *Time Enough For Love* 1973
James Hogan *Thrice Upon a Time* 1980
———— *The Proteus Operation* 1985
Fred Hoyle *October the First is Too Late* 1966
Stephen King *11/22/63* 2011
Damon Knight *Beyond the Barrier* 1964
Michael Kube-McDowell *Alternities* 1988
Henry Kuttner and C. L. Moore (as by Lewis Padgett) "Private Eye" 1949
David Lake *The Man Who Loved Morlocks* 1981
Keith Laumer *Worlds of the Imperium* 1962
———— *The Great Time Machine Hoax* 1964
———— *Dinosaur Beach* 1971
Fritz Leiber *The Big Time, Changewar* 1958, 1983
Murray Leinster "The Runaway Skyscraper" 1919

Ken MacLeod *The Cassini Division* 1998

Elan Mastai *All Our Wrong Todays* 2017

Richard C. Meredith *At the Narrow Passage* 1973

Michael Moorcock *Behold the Man* 1966-69

———— *The Dancers at the End of Time* sequence 1981

C. L. Moore (as by Lawrence O'Donnell) "Vintage Season" 1946

Ward Moore *Bring the Jubilee* 1953

Audrey Niffenegger *The Time Traveler's Wife* 2003

Larry Niven *A World Out of Time* 1976

Claire North *The First Fifteen Lives of Harry August* 2014

Andre Norton *Time Traders* 1958

Andre Norton and Sherwood Smith *Atlantis Endgame* 2002

Joyce Carol Oates *Hazards of Time Travel* 2018

Chad Oliver *Mists of Dawn* 1952

Marge Piercy *Woman On the Edge of Time* 1976

H. Beam Piper Paratime Police sequ. 1948

Doris Piserchia *Mister Justice* 1973

Tim Powers *The Anubis Gates* 1983

Robert Sheckley "A Thief In Time" 1954

T.L. Sherred "E for Effort" 1947

Robert Silverberg *Hawksbill Station* 1968

———— *The Masks of Time/ Vornan-19* 1968

———— *Up the Line* 1969

———— *Project Pendulum* 1987

Clifford Simak *Time and Again* 1951

———— *Ring Around the Sun* 1953

———— *Time is the Simplest Thing* 1961

———— *Our Children's Children* 1974

———— *Mastodonia* 1978

———— *Highway of Eternity* 1986

Vandana Singh "With Fate Conspire" 2013

Michael Swanwick *Bones of the Earth* 2002

Tom Sweterlitsch *The Gone World* 2018

Wilson Tucker *The Lincoln Hunters* 1958

———— *The Year of the Quiet Sun* 1970

A.E. van Vogt "Recruiting Station" aka *Masters of Time*, aka *Earth's Last Fortress* 1942

———— *The Weapon Shops of Isher* 1951

Kurt Vonnegut *The Sirens of Titan* 1959

———— *Slaughterhouse-5* 1969

Ian Watson "The Very Slow Time Machine" 1978

H.G. Wells *The Time Machine* 1895

Charles Williams *Many Dimensions* 1931

Jack Williamson *The Legion of Time* 1938 [rev 1952]

Connie Willis "Fire Watch" 1982

————— *Doomsday Book* 1992

————— *To Say Nothing of the Dog* 1998

————— *Blackout* and *All Clear* 2010

John Wray *The Lost Time Accidents* 2016

John Wyndham "Consider Her Ways" 1956

Charles Yu *How to Live Safely in a Science Fiction Universe* 2010

Some Papers Related to Time Travel Published in Major Scientific Journals

Tipler, Frank J.: "Rotating Cylinders and the Possibility of Global Causality Violation," *Physical Review* D9 (1974: 2203.)

Morris, Michael S., Kim S. Thorne and Ulvi Yurtsever: "Wormholes, Time Machines, and the Weak Energy Condition," *Physical Review Letters* 61, 1988, 1446-49

Visser, Matt: "Traversable Wormholes: Some Simple Examples." *Physical Review* D39 (1989: 3182-84)

Novikov, I.D.: "An Analysis of the Operation of a Time Machine" (English translation) *Soviet Physics JETP* 68 (Mar 1989 439-43)

—————. "The Time Machine and Self-Consistent Evolutions in Problems with Self-Interaction," Preprint NORDITA-90/38A, Feb 1990

Friedman, John, Michael S. Morris, Igor D. Novikov, Fernando Echeverria, Gunnar Klinkhammer, Kim S. Thorne and Ulvi Yurtsever: "Cauchy Problem in Spacetimes with Closed Timelike Curves," *Physical Review* D42 (1990, 1950-30)

Echeverria, Fernando, Gunnar Klinkhammer and Kim S. Thorne: "Billiard Balls in Wormhole Spacetimes with Closed Timelike Curves," *Physical Review* D44 (1991, 1077-99)

Garfield, David, and Andrew Strominger: "Semiclassical Wheeler Wormhole Production," *Physics Letters* B256, 2 (March 1991)

Mallett, Ronald: "Weak Gravitational Field of the Electromagnetic Radiation and a Ring Laser," *Physics Letters* A269 (2000, 214)

Olum, Ken D., and Allen Everet: "Can a Circulating Light Beam Produce a Time Machine?" *Foundations of Physics Letters* 18 (2005: 379-85)

Appendix: "The Dry Sauvages"
Damien Broderick

In his Springer book *Time Travel Tales* (2017), Paul J. Nahin closed with three sf stories of his own. The first of these, he explained, "was written with the specific goal of illustrating how a trip into the past yet to be initiated could logically influence events in the time traveler's present and future." I thought this was a fine idea for a critical commentator, placing oneself in the firing line. So I have done that here with a previously unpublished time machine tale. A brief exegesis follows the story.

In the rather severe twenty-third century apartment the Board has provided, I hard-boil an egg, shell it, let it cool while I slather yellow flavored transfat-free spread on my whole-wheat bread, cut the naked egg lengthwise and sprinkled the golden yolk with sea salt and pepper, then seal it. You can never be sure where your next snack is coming from, and I get peckish.

I'm naked as a boiled egg myself, and step into the sprayer after rinsing my hands. Yes, I've been called reckless and scofflaw in my time (in many times, to be candid) but I am not a fool. No rampaging alien bacteria and viri for me, no sir. As usual the sprayskin stings as it settles. When the rest of me is dry I sit on a stool and raise the soles of my feet for a final squirt.

"You're done, Snow," the shield dispenser tells me. "You may get dressed now."

"Yeah yeah. Or I could just saunter out like this with my medkit slung over one shoulder."

"Not a good idea. Try that and you will find the aperture locked."

© Springer Nature Switzerland AG 2019
D. Broderick, *The Time Machine Hypothesis*, Science and Fiction,
https://doi.org/10.1007/978-3-030-16178-1

"Just messin' with you." I wink at a random spot on the wall, and get into my medic garb.

In the hall I wait for the chronovator. Going up. Only two fellow passengers, both fellows. They glance at me, then politely away.

"Four thousand seventy-nine, CE," I say. "32nd of the fourth month, one hour past noon."

"I can do that," the chronovator door says. "Once again, Snowflake. This is a formal Customs Announcement. You're going to have to dump the organic produce in my bin."

The other two tighten their lips and give me a reproving glance. One male draws himself back ever so slightly against the wall.

"Forget that," I say. "I'll eat it now." I unwrap my snack, nod apologetically to my temporal companions, and wolf the lot down in four quick chomps. I dump the container into the chronovator's disposal. "Yum, they don't know what they're missing."

"Your breath stinks of sulfur," the door says, and extrudes a mug of tasteless but cleansing pale green liquid. I quaff it down, gargling as quietly as I can.

"Come on, come on, these poor people are waiting." Always blame someone else, that's my father's motto.

The chronovator lets them out somewhere in the late third millennium, then takes me to the early fifth.

*   *   *

A young woman I assume to be my epidemiological assistant *du jour* has been alerted and waits expectantly at the door. Behind and above her, as in every Aeviternity lobby, are the words of a long-dead poet, T.S. Eliot:

FARE FORWARD, TRAVELERS!

followed by several oracular if admittedly murky lines of verse about escaping into different future lives. Eliot didn't know about escaping *from* future lives already lived. I look down at the young woman and smile, or at least show my teeth.

"Best welcome to 4079, Dr. Gaspard," she says. She doesn't put out her hand and I don't offer to shake it. Some epochs have funny hygiene phobias.

"Snow," I say.

Her long glossy hair is asymmetric; slightly flustered, she runs a hand over the deep purple depilated right side of her scalp.

"Not Snow, Dr. Gaspard, Kesteesh Ah," she says in faintly accented Common, making "Snow" rhyme with "cow." Artlessly, she lifts my right hand, turns it over, and licks the palm. Okay, *not* the sort of person to be upset by a hard-boiled egg on bread, or a handshake, I can tell that immediately. "In this century we use gender-inflected nomenclature. Were you expecting—"

"*I'm* Snow," I say. "Call me Snowflake if you must be formal. Dr. Gaspard is my mother." She looks even more anxious, so I add, "No, see, I'm also Dr. Gaspard, it's just that I don't usually—"

Kesteesh Ah shakes her head ever so slightly, a gesture rather similar to the movement of my two recent chronovator companions. It must be something I do. In fact, I'm lying slightly. My full given name, thanks to my feckless and saccharin foster-parents, is Blossom Precious Snowflake. You can imagine the fun I had as a kid. But hey, if it weren't for the loony doctors Gaspard I'd still be a frozen orphan zygote back in the early 21st. Can't complain. But stick to "Snow" unless you want a poke in the snout.

Kesteesh or possibly Ah goes on, "Were you expecting to be greeted by a factor for Médecins Sans Frontières de Temps? Allow me to apologize most graciously for their absence, but an emergency—"

"Apology accepted, but unnecessary. When you've been in this game a few years you rarely expect anybody to be on time in time."

"Yes, doctor. I have here the update link to your briefing and inload. Do you accept it?"

"Sure." I follow her to the inloading bay, sit down under one of the resonators. Kesteesh Ah does likewise, opposite, closing her eyes. I wait for the green signal, then tap my left upper third molar with the tip of my tongue. Information zips into my short term cache, and rewrites parts of my nervous system.

I hate it, but what can you do? So much knowledge, endlessly updated, filtered, selected and repackaged from the feeds streaming into the secure local servers in each decade for at least the next half million years. Broken into theorized and empirically validated cycles, granted, but absurdly beyond one person's comprehension, even at the coarsest graining. The largest scaled rises and falls of history are evident enough, especially when a massive random disaster like an undiverted asteroid winter sets everything back almost to scratch for tens of thousands of years. But the microscopic clio-levels, where we doctors intervene to reduce human misery as best we can, defeat any attempt to see beyond their borders, to make of their quilt a larger historical panorama. We leave that to the higher level analysts and their powerful data-mining machines.

And what we do receive in these task-defined inloads is already crammed, predigested, partitioned: elements of the local language, detected and scrutinized by the resident AIs and ported into our transvoxen for easier communication with the locals, a quick pseudomemory of available technology, geography, climate, weather, blah blah, some sense of what has been happening there for the last few decades or centuries and when the vectors are headed in probability space for ensuing centuries or even millennia... It's a lot to shove into one modest human brain, even with biophotonic axonal resonance.

Inloading *hurts*. I'd be screaming, if my vocal apparatus were not switched offline for this routine procedure.

You'd think with all our advanced understanding of the nervous system, the genome, the connectome, bioimmunities, placebos and nocebos, that pain would have been banished by now at will. For most purposes, yes, of course, with sufficient warning. And it's universally known that the brain itself in all its folds and convolutions, can't support or detect pain directly. Is there something metaphysical about the brief agony of inloading, then? Is the soul crying out in rebuke at its invasion? Some make that case. I'm fairly sure they're superstitious ninnies, but when the crush clamp circles my head I have to wonder if my skepticism is the most callow refusal to embrace the self-evident.

*   *   *

After a time I rise, head crammed with what feels like useless and pointless knowledge. At least I know our temporal destination, and why we are being directed there: the young girl who will be the Prophetess if we manage to cure her, is deathly sick, and her whole family and clan are at the edge of starvation and illness. Another working day saving history from itself. The lede directs me—or rather us, since the epidemiologist will be coming with—to the Kairos center of operations. Easy to locate, since almost every Aeviternity time station is built on exactly the same architectural plan. We are directed to Shunt 137, a large boxy instrument prepacked with small sturdy boxes of drugs, medical ancillaries on the order of injector sprays and surgical lasers, solar power shells in condensed mode, quite a lot of tasteless food (don't want the natives getting fixated on the free lunches) and some emergency water, and machines dressed as veiled and cloaked devotees to do the heavy lifting. We take our seats and I activate the shunt. Almost immediately the front of the box slides up and back and searing light smashes against us. I'm grateful, as always, for our Total Protection Garb. Kesteesh Ah rises gracefully from her chair like a silver flame. My own hands look molten, like mercury.

"Welcome to century 137, location largely isomorphic to central Poland in datum century 21," a voice says behind my ears. "External temperature currently Celsius 66.7 degrees, Fahrenheit 152. Do you desire a sensory feed?"

"Yeah, but keep the ambient heat down to a slow roast."

I stagger for a moment as the heat assaults me, filtered though it is by the TPG. For century upon century before my own time, this district has been deeply watered pastureland, abundant grains, cattle, swine, fowl. Now it is close to Saharan. Far to the north, the Arctic ice has long been gone. Here, wretched tents sag. Shorthaired animals cower under ragged covering supported by dry sticks and beaten razor wire. Several warriors emerge, clad like our machines but in grimy white, holding curved blades and one worn energy weapon that probably packs about enough punch to startle a grasshopper. That man steps forward. He is perhaps five and a half feet tall, brutally undernourished from birth, I guess. The fellow stares at our large box and spits into the dried grass. Not cowed, okay.

"You are the Time Angels of Life and Death," he says. "We know your legend." The transvox works well enough in translating his gutturals, "I am Hetman Kee You Wits." A lexical display flashes up: *Kiewicz.*

"Greetings, Hetman. We are here to aid you and your people."

I breathe the parched air through my mouth, and the taste is horrible. I force myself to take a sniff, and regret it.

The place stinks.

You'd think it would be sterilized by the vicious glare, but the hothouse sun cannot compete with the sick, rotting bodies under those patched tents, with the stench of vomit and diarrhea.

"Get them some water. Quick!"

The machines back into the stock of provisions, wheel or carry cool pure water in heavy paper boxes. You can see the desperation in the locals' eyes, the twitching of their hands, but they are not stupid. The Hetman stands aside, watching them vigilantly. The warriors wait for the water to be deployed, calling in their guttural voices to women who bring out drinking vessels, jugs, bowls. Water sparkles. They are careful with it, drinking slowly, sharing. Nobody is foolish enough to waste it in washing the dust from their sunburned flesh. From the tents, though, we hear moans and feeble puking. Kids die easily. There are no old people visible, and probably none anywhere in the encampment.

Another glimpse of the future of humankind.

At a quick glance around, it seems certain that they've lost almost all their sustainable technology. It is incredible that one of those wasted, delirious children under a shabby tent will grow up, if we can rescue her fading life, to forge her people into an army of recovery, redemption, rebirth—however cruel and unforgiving that task.

I'm a physician, have been for some years, so I'm supposed to be hardened against the spectacle of suffering. But I stand there like a silver god, beaten by a glare of the actinic light I am spared solely by the knowledge of my culture, so many centuries lost to these poor souls, and start to weep. Then I shake my head, and crush the tears. I begin to laugh, against my will, as one does in extremity.

"'The place stank,'" I wheeze, speaking in my native English the words that had passed through my mind. "Did you ever see that dreadful entertainment, Kesteesh? Of course you didn't, it was centuries before your time."

"What you ask, Snow?"

That's an advance, so we are on first names now. I can taste the tears as they run down my cheeks.

"Old drama. Fishy murder mystery set in Antarctica. *The Plaice Tank*. The fish eat the murdered victim's corpse."

"This is amusing?"

"No. The review in the *Guardian* was. Three words." I pause. She ponders, but she is quick.

"Oh. Very droll in English. 'The play stank.' " If she smiles, I cannot see it behind her silver mask.

"It was shut down the next night. My father wrote and directed it. He was never the same again."

After a moment of silence, Kesteesh said, "Let us proceed with our rounds, Snow. We have children to deal with, and their parents. And the future Prophetess."

"In the dependent epoch, Kesteesh. If she survives." I shut my eyes, scrolling through the scanty data records of this period. The Hippocratic oath is as superseded as Asimov's classic fictional Laws of Robotics. There is no way to avoid doing harm to some, replacing one life for another. Yes, we are tasked specifically with saving that little girl. But I want to cure as many of the poor devils as I can, given our limitations. "The women won't let me near the kids," I say. Of course she knows that. Why do I feel useless? I am far from useless. "I'll look after the adult men and any boys they bring out."

The bright silver statue nods her head, and she goes to intervene in history, to beat against the current, borne forward ceaselessly into the future, yes, to forge the recreated consciousness of our species. Those old poets! Man!

The furnace wind blows across the ruined landscape. How do they manage life here and now, these wild, untamed remnants who are the legacy of the sins of my own time, of all the gorged and poisoned times? Men wrapped against the sun and the khamsin, the sirocco, the harmattan are pushing toward me, these dry sauvages (yes, I remember my studied Eliot) in a parched, savage era, bringing their children with them to my ministrations. I do not weep, this time. I draw out my instruments and set to work.

*   *   *

After the sun sets, the baked surface radiates in the dark and then cools fast, shedding its heat. We are at 52° North, what else can I expect of Greenhouse Catastrophe Earth? We choose to withdraw into our large impregnable box and eat our own food, drink a glass of wine each. Maintain the mystique. The moment we got inside we had shed our TPGs and, naked, sprayed each other clean. I watch Kesteesh with slowly dawning interest. While I am not enthralled by her hair style, she probably doesn't think much of mine either. A machine brings us light robes and dims the lights.

"Shall we sleep together?" I ask.

"Absolutely not."

I am very faintly offended, but she might have a hundred reasons.

"I hope I have not violated your *mores*. Well, shall we have another glass of this admirable wine before we—"

"Snow, I would enjoy sex. But no sleeping in a bed after. I am sure you snore."

Snore? Me? How can she know? The word gets around. I guess. I should have my antrums corrected.

"Let me clean my teeth, dear epidemiologist."

"And I."

We have a nice time, and so to bed. Separately.

*   *   *

After eight days, we've done all we can for these poor people. What other option? Vaccinate the entire planet? Feed the remnants of humanity in this epoch with the diminishing resources of earlier centuries? That is a utopian absurdity, and one leading straight into blindness. Tweak or shove the nonlinear dynamics of the long cycles, and everything is liable to unravel into dystopia. Kesteesh and I have wasted hours after dark chewing this old bone, as every time medic tends to do, without getting anywhere. Finally our tour of duty is done. Briefly and without ceremony we farewell Hetman Kiewicz and his lieutenants. Of necessity we ignore the little girl whose life we have come to spare, although Kesteesh admits she's given her a special farewell hug. The machines haul back all the anomalous materials we brought from the past, button up the box, and implode back to the same second we left. Frankly, I just want to sign out and go back to my twenty-third century apartment. I have become rather fond of Kesteesh Ah but there is no future, quite literally, in that. I hoist my medkit on one shoulder, shake her hand (she's got past that particular phobia or local custom), and a feed seems to whistle behind my ear.

"That would be the emergency," Kesteesh says.

"Do you remember what it was? Is?"

"Full staff meeting in the auditorium, is all I recall." The feed in my ear doesn't tell any more than that.

I sigh. Les nouvelles Annalistes Doctrine update, probably. Always fun.

"Let's toddle."

*   *   *

As we walk down the corridor a bulgy specimen lurches up out of a low slung visitor's seat, takes three thudding steps, and pushes his face into mine. I draw back in disbelief. This is just plain rude.

"Hey, you bad," he says in a raspy, emotional voice at the edge of breaking, "you're the badding bad Snowflake Gaspard, aren't you?"

I back away another step and scrutinize his features and other bodily details. Overweight certainly (and how has he managed *that*, in this place and time, with the excellent dietetics we medicos abide by?), heavy shoulders, bloodshot green eyes, furious mouth. I've never seen him before in my life.

"I'm Snow," I say. "What can I—?"

The blow doesn't quite break my nose but it hurts atrociously.

"Ow! What the *bad*?" I yelp, falling over backwards. A security field catches me, lifts me back on my feet. I crouch. The man's other fist passes through the place I've been. Some distant part of my attention hears Kesteesh Ah mutter "*Contretemps* alert." Instantly, a red light is flashing. Two women come quickly into the hall behind the thug and drop a shimmery net over him. I've never see that done before, either, although of course I'm familiar with the technique. The net clamps him tightly, beefy arms pulled in tight against his chest, leaving his calves unbound. They guide him hobbling and hopping back to his seat and apply a chemical spray with a sweet scent of roses, softening just the salient areas of the net that let him sit, and then a follow-up spray that holds him tight again.

"Sir," says one of the women, whose hair style rather resembles Kesteesh Ah's asymmetry, "you must behave. This is a medical facility, not a sports rink."

"It's a godbad nest of murderers like this badding very bad Gaspard."

(Please forgive me for all this "bad" bullbad. The Common transvox system in this era is stupidly puritanical and mealymouthed.)

I take the seat next to his, frowning.

"Murderers? You must have me confused with Jack the—"

"You murdered my wife Aspandrel. You motherbad, you never gave her chance to defend herself. You deleted her from history!"

Oh. Oh, bad!

"Pardon me, doctor," one of the monitors says, spraying a silencer over the lips of the pugilist then probing my nose with an insufflator that squirts cool and healing moisture into my nostrils. "You cannot communicate with your assailant. As soon as you are comfortable, please depart."

I palm my nasal cartilage. Not fractured. Can't feel a thing. In a somewhat muffled voice, I say, "On our way, *tout de suite.*"

*   *   *

By the time Kesteesh and I reach the auditorium, I've tried to access the lunatic's bio and log and been told sharply that this information, including his name, is interdicted. Doesn't take much to put two and two together, but that fails to get me anywhere useful. Obviously he is from a temporal epoch I haven't yet visited but am going to. There might be no such thing as a paradox, but *la longue durée* operatives have too much at stake to allow the slightest risk. I will meet up with him sooner or later, and if they don't expunge my memories of this little incident I'll discover whatever I am going to do to upset him frightfully. The accusation of murder is ludicrous and, really, rather offensive. The question is why the staff allowed him into the spacetime where we had our unfortunate meeting. It must have been deliberate, at some level way above my grade. Letting him blow off steam at my expense? For the moment I put it out of my mind and follow Kesteesh into the dimmed room.

The presentation is tedious. I lean over and muttered to Kesteesh, "Ideological claptrap. If it's not one claque it's another."

"Shoosh," she tells me. "If you paid more attention, Snow Gaspard, you wouldn't get into fights with people."

"Fights?" I have no idea what she's talking about. Something is wrong with my nose. I snuffle, and it feels blocked. Oh. Blocked is right. Something happened recently and the system has shut it out of my memory access. Stupid idea. How can you learn from your mistakes, or other people's mistakes, if you're prevented from knowing what they were. Maybe Kesteesh Ah could tell me but then they've doubtless blanked her on the topic as well. I subside into my seat and try to pay attention.

The notification claimed an emergency, but it seems just the usual squabble between Hippocratic healers (like me and my epidemiologist) and sealers (who favored the broad brush approach). I see other smaller clots of partisan groupules: the mockingly named wheelers and dealers, who want to open markets between the eons and are lucky they aren't all in prison, the peelers who fancy a corps of Time Police, a few others whose political positions do not lend themselves to such natty rhymes. Generally the lesser fry keep their heads down and their voices muted, although I see them flicking messages back and forth as the Annalistes Commissionner Jean-Marie Backster argues his version of the classic *mentalities* model of epochal psychologies. In a moment, or an hour, the Honorable Omuyu Akerele will call for leave to rebut these arguments with copious evidence drawn from the Kairological doctrine, rich contextual thick descriptions so much more fluent and relevant, in that doctrinal view, than the scientistic chronological clock-tocking of the Annales party.

My nose is starting to itch, and it seems to the touch to have swollen. I want nothing more than to get out of the room and go home after this week's work, but I'll be fined if I leave before this mutually undermining propaganda is completed.

"Wake up!" my companion growls in my ear. "You're snoring! I knew you were a snorer."

"Urp? Have they got to the crisis yet?"

"Just started, I think."

Professor Akerele stands at the front of the auditorium, tall, broad-shouldered, dark as a carved mask. Set-theoretic calculus covers the display area at his back, echoed in my left eye display until I blink it off.

"Colleagues, let me sum up. In essence, and indeed in hoary cliché, two major potential actions are available to us." The groupules shift resentfully, excluded by this binary contrast, but keep their voices low. "Kairos holds attention fixed on individuals and their affinities across humanly meaningful generations. Annales looks instead to the great waves and cycles of vigor, torpor, invention and degeneration, intellect versus passion or resentment—" He breaks off as Backster repudiates this simplified parody. "Sir, please—My honorable colleague has spoken already at length, asserting the benefits of very long durational analysis. I am here today, perhaps to his surprise and that of many delegates and operatives, to agree with him."

Now there is an audible rustling and several cries of outrage. A system mutes these objections somewhat, and raises Akrele's voice a little. It is enough.

"We have come at last to the discontinuity both our observational data and our converging hypotheses have warned us of—or arguably promised us—that lay ahead all these centuries."

"Sir, " Backster objects, "we are few in number, with comparatively limited funding, and that evidence gathered by this small institute and our machines is scant to the point of guesswork. Yes, we have now a rudimentary mapping of the great durational epoches and epistemes, but this is still very far from the necessary fine-grained detail we need to allow—"

"Sir, permit me to make my presentation, and then we can turn to a discussion of what must yet be done to renormalize history. To begin, we agree on the basic structure of the quantum relativistic evolving block universe. Time is the boundary membrane of an expanding, foaming moment experienced as the present by whatever observers are manifested there. So the future is open at every moment, from the standpoint of that moment. The future is thus infinitely although not transfinitely emergent. The past is rigid, in that globular block, *from the perspective of each observer.*"

"Who questions these ancient certainties ? They provide both the miseries of the long cycles and our determination to mitigate those sorrows within the severe bounds of probability."

"Indeed yes, my good colleague. Let us take the next step into a practical corollary only dimly glimpsed until now, in the beginning of the fifth millennium. I call Dr. Kesteesh Ah to stand and address the conclave."

What? Surely not two operatives here with that same odd name—

My recent lover and medical partner rises and the machineries manifest her image between the two rival ideologues.

A virtual space opens behind her head, displaying the sorry village of starving, sickened people we ministered to, within the limits permitted us by the laws of the Temporal Authority. Or was it so? A sickening cramp seizes at my innards, and a mortifying sense of shame. That child! That doomed little girl who in the dominant probability history had died, with most of the rest of her small remnant tribe. Now she lives, or will, in the future we'd refurbished. She will grow under the whip of her ferocious vitality into an adult who becomes the messiah, for a 1000 years, of a dying Earth. Her hard, numinous message of hope and punishment will flog her people toward a recovery of knowledge. The surface of the evolving block universe will buckle and readjust (has already done so, surely!), making a new history from the ruinous exhaustion that has held future humankind on a course to inanition and finally extinction.

This memory crashes into my soul like an inload, but where an inload hurts the brain and brings forth throttled, stilled screams, this tears at my very self worth, my delusion of agency, my edited reality.

Kesteesh Ah is speaking with a measure of gravity I never noticed in her conversations, however technical, however charmingly akin to pillow talk without benefit of the pillow.

"With the aid of my colleague. Dr. Blossom Precious Snowflake Gaspard," and luckily I hear not a titter of stifled mirth at my name, "and drawing upon the epidemiological data from caches maintained across ten millennia beginning with year 171,000, I established a seven sigma probability that any future derived from the so-called Prophetess would terminate in local catastrophe. This provided a suitable contained workspace for the model currently under study by the consortium. In line with that program, and its ethics oversight committee, I injected all the females of the remnant tribe intravenously with the barbiturate sodium thiopental, augmented by intramuscular midazolam and hydromorphone, bringing peaceful and painless demise within minutes. Meanwhile, Dr. Gaspard, who was blinded to the cliotropic purpose of this experiment, had treated the males with the same specifics. Any objection on ethical grounds should be directed entirely to me and the principal investigators."

I am dazed, unable to accept what she has said, what I think she has said. Impossible. The muted muttering in the auditorium blows up into protesting shouts, noisy discussion. The woman sitting on my other side stares furiously at us. Kesteesh Ah is emotionally unmoved, or so it seems.

"Let us now consider the implications for using this workspace to test the limits of resetting the *mentalités* of entire epochs, fine-tuning the psychological dominants of the long durations. Finally," and she raises her voice, "we are truly operating in the aevum—this temporal space between the supposed timelessness of a deity, and the rigid world-lines of quantum relativity."

*   *   *

I get to my feet without looking at her again, push past the furious woman and others in my row, stumble to the exit. Without exaggerating, I have to say that I am close to throwing up all over the carpet. Machines and fields would clean up the mess, but they are not subtle enough to cleanse my horrified self-disgust. I wipe my eyes and blow my nose, which is strangely tender and still feels swollen, as if I've banged my face hard on something solid.

A machine finds me crouched on my heels in the corridor, doubled over, and hoists me to my feet.

"It's not the dead little girl," I try to explain to the machine. "I mean, it is, of course. And the rest of her poor tribe."

"Come this way, Dr. Gaspard. You have been under a lot of stress." It leads me into a small room and helps me onto a comfortable pallet, removing my shoes and covering me with a light blanket. Something else like a light blanket moves across my brain, makes my eyes twitch. Bad bad bad. They are reinitializing my memory.

"Not just the girl," I say, before the warm dark takes me away. "All the other lives, a thousand years thirty thousand forever, snuffed out. I know there'll be new people. They'll be happier, I suppose. Oh *bad*!"

Is that the point? Why I was part of the experiment?

Good night, good night, good night

* * *

I snatch a hasty evening meal at a commercial Tomorrow Port golden arches booth, clean my teeth in their sparkling restroom, then take the chronovator to springtime, 22,341. As in every Aeviternity lobby I find posted the words of the long-dead poet, T.S. Eliot:

FARE FORWARD, TRAVELERS!

followed by several oracular if admittedly murky lines of verse about escaping into different future lives. Eliot didn't know about escaping *from* future lives already lived and shortly to be redacted.

I have traveled without a pill and emerge into the realtime environment at noon on April 22 (by my ancient calendar, anyway), which in Arctica is warm, wet, bright as high summer. Should have taken the melatonin after all. I am instantly timelagged, that slightly sickening drained sensation all through my flesh, but leave my shades in a top pocket and keep my bleary eyes open. Sunlight resets your circadian rhythms, if you let it, and there is plenty of that here and now.

Lucky, really. If I'd shut my eyes, the man I am watching from the corner of my eye, rather than via my detectable sensor feed, might have wandered off into the dismal, cluttered marketplace. I can't put my finger on what has alerted me to him. Is he a fellow aeviternal, perhaps a Kairo? He is unpleasantly bulgy, by the standards of this dilapidated and underfed century, heavy shoulders, wild green eyes, mouth puckered to kiss the short, faded woman, dressed like a local, who displays every indication of sorrowful parting disguised as desperate good cheer. I can't imagine what she sees in him. He catches me glancing at them both, gives me a goofy grin and a wink. Urp. Not so good. He is obviously a time courier, and almost certainly he has no idea that Ah and I are about to expunge and upgrade this entire sheaf. For the common good, you know?

I find Kesteesh Ah regarding a large hairy dog in a field container, barking its head off.

"Oh no," I cry, and touch her shoulder. "Timmy's in the well!" I have to drop out of Common into my home tongue to utter that old phrase. Silly, but it makes me smile.

"Dr. Snow Gaspard, how are you today?" Briefly and lightly, she licked my palm. "So we're speaking English today. What's a 'well'? Who or what is 'Timmy'?"

"A well is a hole in the ground with water at the bottom, Dr. Ah, and Timmy is a small rather stupid boy who has climbed into it and can't get out."

"So the dog is upset and runs to alert Timmy's proctor."

"Close enough." I study the animal in the field container. It is licking its bad. "I grant you, this canine specimen doesn't look intelligent enough."

"Timmy's in the Bell," she says speculatively.

"No, no, you can't be *in* a bell. That'd be upside down."

"Timmy's in the Dell."

"Great old English word! But that wouldn't upset Lassie. Lassie's the guard dog."

"Timmy's in the Cell. Timmy's in the Jail." She pronounced "jail" as if it rhymed with "gell," as it probably does when she comes from. "Timmy was Sellin', Tellin' and Yellin' and he Fell In."

I am grinning. Gotta love that Gal. Within the hour, we will be reconfiguring this future, which is in worse shape than Timmy's imaginary anodyne world by many increments. Ah carries a potent virus vector in her kit, ready to inject into Lassie before setting her free. The locals have probably forgotten how to dig wells. We have ways to cure that, though, to revive their science and technology, we angelic aeviternals. I rub my nose, which suddenly itches. Do I really snore?

It strikes me that Kesteesh Ah looks tired. Cafard, they once called it. Melancholia. Weariness of the dispirited spirit. But somebody has to do it. On the right side of her scalp, her hair is growing out blonde.

She bends away from me to pick up her medkit. "Timmy's in Hell," she says.

A Brief Exegesis

My fascination with time travel probably began with the newspaper serial cartoon *Brick Bradford*, a mid-century adventurer who flew through time as well as space in his lightbulb-shaped Time Top. Soon I was dazzled by sf stories such as A.E. van Vogt's *The Weapon Shops of Isher* (1951), in which an unlucky fellow was flung back and forth through time, accumulating a monstrous temporal charge and ending with its titanic, explosive release: "He would not witness but he would aid in the formation of the planets."

After that astonishing act of cosmic invention, other time travel stories moved into the realms of the sportive, the *policier* and even the administrative (Asimov's *The End of Eternity* was a prime example of that). Sf's bent had always tended to the "idea as hero"; my first serious novel, *The Dreaming Dragons* (1980), introduced the generation time machine, the temporal equivalent of science fiction's many-generation starship for extremely lengthy journeys.

By the twenty-first century, though, rich characterization strengthened these thought experiments with greater realism and many more pages to express it (*The Time Traveler's Wife*, for example). Still, sf has always found a special place for the focused ingenuities of the short story or novella. I set out with this short piece to craft a waggish tale that drew upon a variety of classic time travel tropes: interventions in history that paid off in both personal and long-cultural terms; the ready use of brain technologies to protect time operatives from the dire consequences of their necessary actions; bureaucratization of the ineffable; compartmentalization of the intolerable; leaks in the system both deliberate and accidental.

Oh, and I chose deliberately to use that device much reviled by writing school instructors, the "Tell Don't Just Show" crime of talking heads swapping compressed information—which here in the real world is actually the way we communicate and learn almost everything of importance to a species that talks and thinks and hopes. Me, I hope that readers will have gone with Snow's flow. He's an unreliable narrator, with some of that unreliability imposed on him, more than once, but then so are we all, even those of us without the benefit and curse of time machines.

Index

© Springer Nature Switzerland AG 2019
D. Broderick, *The Time Machine Hypothesis*, Science and Fiction,
https://doi.org/10.1007/978-3-030-16178-1

Printed in the United States
By Bookmasters